T0314184

ORGANIC SYNTHESES

ORGANIC SYNTHESES

AN ANNUAL PUBLICATION OF SATISFACTORY METHODS FOR THE PREPARATION OF ORGANIC CHEMICALS

VOLUME 82

2005

DEPARTMENT OF CHEMISTRY, UNIVERSITY OF NEW HAMPSHIRE
DURHAM, NEW HAMPSHIRE 03824

A JOHN WILEY & SONS, INC., PUBLICATION

The procedures in this text are intended for use only by persons with prior training in the field of organic chemistry. In the checking and editing of these procedures, every effort has been made to identify potentially hazardous steps and to eliminate as much as possible the handling of potentially dangerous materials; safety precautions have been inserted where appropriate. If performed with the materials and equipment specified, in careful accordance with the instructions and methods in this text, the Editors believe the procedures to be very useful tools. However, these procedures must be conducted at one's own risk. Organic Syntheses, Inc., its Editors, who act as checkers, and its Board of Directors do not warrant or guarantee the safety of individuals using these procedures and hereby disclaim any liability for any injuries or damages claimed to have resulted from or related in any way to the procedures herein.

Published by John Wiley & Sons, Inc., Hoboken, New Jersey.

Published simultaneously in Canada.

"John Wiley & Sons, Inc. is pleased to publish this volume of Organic Syntheses on behalf of Organic Syntheses, Inc. Although Organic Syntheses, Inc. has assured us that each preparation contained in this volume has been checked in an independent laboratory and that any hazards that were uncovered are clearly set forth in the write-up of each preparation, John Wiley & Sons, Inc. does not warrant the preparations against any safety hazards and assumes no liability with respect to the use of the preparations."

Library of Congress Catalog Card Number: 21-17747
ISBN-13 978-0-471-68256-1
ISBN-10 0-471-68256-X

Printed in the United States of America
10 9 8 7 6 5 4 3 2 1

ORGANIC SYNTHESES

VOLUME	VOLUME EDITOR	PAGES
I*	†ROGER ADAMS	84
II*	†JAMES BRYANT CONANT	100
III*	†HANS THACHER CLARKE	105
IV*	†OLIVER KAMM	89
V*	†CARL SHIPP MARVEL	110
VI*	†HENRY GILMAN	120
VII*	†FRANK C. WHITMORE	105
VIII*	†ROGER ADAMS	139
IX*	†JAMES BRYANT CONANT	108
Collective Vol. I	A revised edition of Annual Volumes I–IX †HENRY GILMAN, *Editor-in-Chief* 2nd Edition revised by †A. H. BLATT	580
X*	†HANS THACHER CLARKE	119
XI*	†CARL SHIPP MARVEL	106
XII*	†FRANK C. WHITMORE	96
XIII*	†WALLACE H. CAROTHERS	119
XIV*	†WILLIAM W. HARTMAN	100
XV*	†CARL R. NOLLER	104
XVI*	†JOHN R. JOHNSON	104
XVII*	†L. F. FIESER	112
XVIII*	†REYNOLD C. FUSON	103
XIX*	†JOHN R. JOHNSON	105
Collective Vol. II	A revised edition of Annual Volumes X–XIX †A. H. BLATT, *Editor-in-Chief*	654
20*	†CHARLES F. H. ALLEN	113
21*	†NATHAN L. DRAKE	120
22*	†LEE IRVIN SMITH	114
23*	†LEE IRVIN SMITH	124
24*	†NATHAN L. DRAKE	119
25*	†WERNER E. BACHMANN	120
26*	†HOMER ADKINS	124
27*	†R. L. SHRINER	121
28*	†H. R. SNYDER	121
29*	†CLIFF S. HAMILTON	119
Collective Vol. III	A revised edition of Annual Volumes 20–29 †E. C. HORNING, *Editor-in-Chief*	890
30*	†ARTHUR C. COPE	115
31*	†R. S. SCHREIBER	122

**Out of print.*
†Deceased.

VOLUME	VOLUME EDITOR	PAGES
32*	†RICHARD T. ARNOLD	119
33*	†CHARLES C. PRICE	115
34*	†WILLIAM S. JOHNSON	121
35*	†T. L. CAIRNS	122
36*	N. J. LEONARD	120
37*	†JAMES CASON	109
38*	†JOHN C. SHEEHAN	120
39*	†MAX TISHLER	114
Collective Vol. IV	A revised edition of Annual Volumes 30–39 NORMAN RABJOHN, *Editor-in-Chief*	1036
40*	†MELVIN S. NEWMAN	114
41*	JOHN D. ROBERTS	118
42*	†VIRGIL BOEKELHEIDE	118
43*	†B. C. MCKUSICK	124
44*	†WILLIAM E. PARHAM	131
45*	†WILLIAM G. DAUBEN	118
46*	E. J. COREY	146
47*	†WILLIAM D. EMMONS	140
48*	†PETER YATES	164
49*	KENNETH B. WIBERG	124
Collective Vol. V	A revised edition of Annual Volumes 40–49 HENRY E. BAUMGARTEN, *Editor-in-Chief*	1234
Cumulative Indices to Collective Volumes, I, II, III, IV, V	†RALPH L. AND †RACHEL H. SHRINER, *Editors*	
50*	RONALD BRESLOW	136
51*	†RICHARD E. BENSON	209
52*	HERBERT O. HOUSE	192
53*	ARNOLD BROSSI	193
54*	ROBERT E. IRELAND	155
55*	†SATORU MASAMUNE	150
56*	†GEORGE H. BÜCHI	144
57*	CARL R. JOHNSON	135
58*	†WILLIAM A. SHEPPARD	216
59*	ROBERT M. COATES	267
Collective Vol. VI	A revised edition of Annual Volumes 50–59 WAYLAND E. NOLAND, *Editor-in-Chief*	1208
60*	†ORVILLE L. CHAPMAN	140
61*	†ROBERT V. STEVENS	165

**Out of print.*
†*Deceased.*

**Out of print.*
†*Deceased.*

VOLUME	VOLUME EDITOR	PAGES
80	Steven Wolff	259
81	Rick L. Danheiser	296
82	Edward J. J. Grabowski	225

Collective Volumes, Collective Indices to Collective Volumes I–IX, Annual Volumes 75–81 and Reaction Guide are available from John Wiley & Sons, Inc.

* *Out of print.*
† *Deceased.*

NOTICE

With Volume 62, the Editors of *Organic Syntheses* began a new presentation and distribution policy to shorten the time between submission and appearance of an accepted procedure. The soft cover edition of this volume is produced by a rapid and inexpensive process, and is sent at no charge to members of the Organic Division of the American Chemical Society, Gesellchaft Deutscher Chemiker, Polskie Towarzystwo Chemiczne, Royal Society of Chemistry, and The Society of Synthetic Organic Chemistry, Japan. The soft cover edition is intended as the personal copy of the owner and is not for library use. The hard cover edition is published by John Wiley and Sons, Inc., in the traditional format, and it differs in content primarily by the inclusion of an index. The hard cover edition is intended primarily for library collections and is available for purchase through the publisher. Incorporation of graphical abstracts into the Table of Contents began with Volume 77. Annual volumes 70-74 and 75-79 have been incorporated into five-year versions of the collective volumes of *Organic Syntheses* that appeared as Collective Volume IX and X in the traditional hard cover format, available for purchase from the publishers. The Editors hope that the new Collective Volume series, appearing twice as frequently as the previous decennial volumes, will provide a permanent and timely edition of the procedures for personal and institutional libraries. The Editors welcome comments and suggestions from users concerning the new editions.

Organic Syntheses, Inc., joined the age of electronic publication in 2001 with the release of its free web site (www.orgsyn.org) followed in 2003 with the completion of a commercially available electronic database (www.mrw.interscience.wiley.com/osdb). Organic Syntheses, Inc., fully funded the creation of the free website at www.orgsyn.org in a partnership with CambridgeSoft Corporation and Data-Trace Publishing Company. The success of this site in its first full year of operation was overwhelming, with an average of nearly 48,000 site hits/day and more than 27,000 page views/day. The site is accessible to most internet browsers using Macintosh and Windows operating systems and may be used with or without a ChemDraw plugin. Because of continually evolving system requirements, users should review software compatibility at the website prior to use. John Wiley & Sons, Inc., and Accelrys, Inc., partnered with Organic Syntheses, Inc., to develop the new database (www.mrw.interscience.wiley.com/osdb) that is available for license with internet solutions from John Wiley & Sons, Inc. and intranet solutions from Accelrys, Inc.

Both the commercial database and the free website contain all annual and collective volumes and indices of *Organic Syntheses*. Chemists can draw structural queries and combine structural or reaction transformation queries with full-text and bibliographic search terms, such as chemical name, reagents, molecular formula, apparatus, or even hazard warnings or phrases. The preparations are categorized into reaction types, allowing search by category. The contents of individual or collective volumes can be browsed by lists of titles, submitters' names, and volume and page references, with or without reaction equations.

The commercial database at www.mrw.interscience.wiley.com/osdb also enables the user to choose his/her preferred chemical drawing package, or to utilize several freely available plug-ins for entering queries. The user is also able to cut and paste existing structures and reactions directly into the structure search query or their preferred chemistry editor, streamlining workflow. Additionally, this database contains links to the full text of primary literature references via CrossRef, ChemPort, Medline, and ISI Web of Science. Links to local holdings for institutions using open url technology can also be enabled. The database user can limit his/her search to, or order the search results by, such factors as reaction type, percentage yield, temperature, and publication date, and can create a customized table of reactions for comparison. Connections to other Wiley references are currently made via text search, with cross-product structure and reaction searching to be added in the coming year. Incorporations of new preparations will occur as new material becomes available.

INFORMATION FOR AUTHORS OF PROCEDURES

Organic Syntheses welcomes and encourages submissions of experimental procedures that lead to compounds of wide interest or that illustrate important new developments in methodology. Proposals for *Organic Syntheses* procedures will be considered by the Editorial Board upon receipt of an outline proposal as described below. A full procedure will then be invited for those proposals determined to be of sufficient interest. These full procedures will be evaluated by the Editorial Board, and if approved, assigned to a member of the Board for checking. In order for a procedure to be accepted for publication, each reaction must be successfully repeated in the laboratory of a member of the Editorial Board at least twice, with similar yields (generally ±5%) and selectivity to that reported by the submitters.

Organic Syntheses Proposals

A cover sheet should be included providing full contact information for the principal author and including a scheme outlining the proposed reactions (an Organic Syntheses Proposal Cover Sheet can be downloaded from the *Organic Syntheses* websites). Attach an outline proposal describing the utility of the methodology and/or the usefulness of the product. Identify and reference the best current alternatives. For each step, indicate the proposed scale, yield, method of isolation and purification, and how the purity of the product is determined. Describe any unusual apparatus or techniques required, and any special hazards associated with the procedure. Identify the source of starting materials. Enclose copies of relevant publications (attach pdf files if an electronic submission is used).

Submit proposals by mail or as email attachments to:

Professor Charles K. Zercher
Associate Editor, *Organic Syntheses*
Department of Chemistry
University of New Hampshire
23 College Road, Parsons Hall
Durham, NH 03824

For electronic submissions: *org.syn@unh.edu*

Submission of Procedures

Authors invited by the Editorial Board to submit full procedures should prepare their manuscripts in accord with the Instructions to Authors which may be obtained from the Associate Editor or downloaded from the *Organic Syntheses* websites. Submitters are also encouraged to consult earlier volumes of *Organic Syntheses* for models with regard to style, format, and the level of experimental detail expected in *Organic Syntheses* procedures. Manuscripts should be submitted in triplicate to the Associate Editor. Electronic submissions are encouraged; procedures will be accepted as e-mail attachments in the form of Microsoft Word files with all schemes and graphics also sent separately as ChemDraw files.

Procedures that do not conform to the Instructions to Authors with regard to experimental style and detail will be returned to authors for correction. Authors will be notified when their manuscript is approved for checking by the Editorial Board, and it is the goal of the Board to complete the checking of procedures within a period of no more than six months.

Additions, corrections, and improvements to the preparations previously published are welcomed; these should be directed to the Associate Editor. However, checking of such improvements will only be undertaken when new methodology is involved. Substantially improved procedures have been included in the Collective Volumes in place of a previously published procedure.

NOMENCLATURE

Both common and systematic names of compounds are used throughout this volume, depending on which the Volume Editor felt was more appropriate. The Chemical Abstracts indexing name for each title compound, if it differs from the title name, is given as a subtitle. Systematic Chemical Abstracts nomenclature, used in the Collective Indexes for the title compound and a selection of other compounds mentioned in the procedure, is provided in an appendix at the end of each preparation. Chemical Abstracts Registry numbers, which are useful in computer searching and identification, are also provided in these appendices. Whenever two names are concurrently in use and one name is the correct Chemical Abstracts name, that name is preferred.

ACKNOWLEDGMENT

Organic Syntheses wishes to acknowledge the contributions of Merck & Co. and Pfizer, Inc. to the success of this enterprise through their support, in the form of time and expenses, of members of the Boards of Directors and Editors.

HANDLING HAZARDOUS CHEMICALS
A Brief Introduction

General Reference: *Prudent Practices in the Laboratory*; National Academy Press; Washington, DC, 1995.

Physical Hazards

Fire. Avoid open flames by use of electric heaters. Limit the quantity of flammable liquids stored in the laboratory. Motors should be of the nonsparking induction type.

Explosion. Use shielding when working with explosive classes such as acetylides, azides, ozonides, and peroxides. Peroxidizable substances such as ethers and alkenes, when stored for a long time, should be tested for peroxides before use. Only sparkless "flammable storage" refrigerators should be used in laboratories.

Electric Shock. Use 3-prong grounded electrical equipment if possible.

Chemical Hazards

Because all chemicals are toxic under some conditions, and relatively few have been thoroughly tested, it is good strategy to minimize exposure to all chemicals. In practice this means having a good, properly installed hood; checking its performance periodically; using it properly; carrying out all operations in the hood; protecting the eyes; and, since many chemicals can penetrate the skin, avoiding skin contact by use of gloves and other protective clothing at all times.

a. Acute Effects. These effects occur soon after exposure. The effects include burn, inflammation, allergic responses, damage to the eyes, lungs, or nervous system (e.g., dizziness), and unconsciousness or death (as from overexposure to HCN). The effect and its cause are usually obvious and so are the methods to prevent it. They generally arise from inhalation or skin contact, so should not be a problem if one follows the admonition "work in a hood and keep chemicals off your hands".

Ingestion is a rare route, being generally the result of eating in the laboratory or not washing hands before eating.

b. Chronic Effects. These effects occur after a long period of exposure or after a long latency period and may show up in any of numerous organs. Of the chronic effects of chemicals, cancer has received the most attention lately. Several dozen chemicals have been demonstrated to be carcinogenic in man and hundreds to be carcinogenic to animals. Although there is no simple correlation between carcinogenicity in animals and in man, there is little doubt that a significant proportion of the chemicals used in laboratories have some potential for carcinogenicity in man. For this and other reasons, chemists should employ good practices at all times.

The key to safe handling of chemicals is a good, properly installed hood, and the referenced book devotes many pages to hoods and ventilation. It recommends that in a laboratory where people spend much of their time working with chemicals there should be a hood for each two people, and each should have at least 2.5 linear feet (0.75 meter) of working space at it. Hoods are more than just devices to keep undesirable vapors from the laboratory atmosphere. When closed they provide a protective barrier between chemists and chemical operations, and they are a good containment device for spills. Portable shields can be a useful supplement to hoods, or can be an alternative for hazards of limited severity, e.g., for small-scale operations with oxidizing or explosive chemicals.

Specialized equipment can minimize exposure to the hazards of laboratory operations. Impact resistant safety glasses are basic equipment and should be worn at all times. They may be supplemented by face shields or goggles for particular operations, such as pouring corrosive liquids. Because skin contact with chemicals can lead to skin irritation or sensitization or, through absorption, to effects on internal organs, protective gloves should be worn at all times.

Laboratories should have fire extinguishers and safety showers. Respirators should be available for emergencies. Emergency equipment should be kept in a central location and must be inspected periodically.

MSDS (Materials Safety Data Sheets) sheets are available from the suppliers of commercially available reagents, solvents, and other chemical materials; anyone performing an experiment should check these data sheets before initiating an experiment to learn of any specific hazards associated with the chemicals being used in that experiment.

DISPOSAL OF CHEMICAL WASTE

General Reference: *Prudent Practices in the Laboratory* National Academy Press, Washington, D.C. 1996

Effluents from synthetic organic chemistry fall into the following categories:

1. **Gases**
   - 1a. Gaseous materials either used or generated in an organic reaction.
   - 1b. Solvent vapors generated in reactions swept with an inert gas and during solvent stripping operations.
   - 1c. Vapors from volatile reagents, intermediates and products.
2. **Liquids**
   - 2a. Waste solvents and solvent solutions of organic solids (see item 3b).
   - 2b. Aqueous layers from reaction work-up containing volatile organic solvents.
   - 2c. Aqueous waste containing non-volatile organic materials.
   - 2d. Aqueous waste containing inorganic materials.
3. **Solids**
   - 3a. Metal salts and other inorganic materials.
   - 3b. Organic residues (tars) and other unwanted organic materials.
   - 3c. Used silica gel, charcoal, filter aids, spent catalysts and the like.

The operation of industrial scale synthetic organic chemistry in an environmentally acceptable manner* requires that all these effluent categories be dealt with properly. In small scale operations in a research or

*An environmentally acceptable manner may be defined as being both in compliance with all relevant state and federal environmental regulations *and* in accord with the common sense and good judgement of an environmentally aware professional.

academic setting, provision should be made for dealing with the more environmentally offensive categories.

1a. Gaseous materials that are toxic or noxious, e.g., halogens, hydrogen halides, hydrogen sulfide, ammonia, hydrogen cyanide, phosphine, nitrogen oxides, metal carbonyls, and the like.
1c. Vapors from noxious volatile organic compounds, e.g., mercaptans, sulfides, volatile amines, acrolein, acrylates, and the like.
2a. All waste solvents and solvent solutions of organic waste.
2c. Aqueous waste containing dissolved organic material known to be toxic.
2d. Aqueous waste containing dissolved inorganic material known to be toxic, particularly compounds of metals such as arsenic, beryllium, chromium, lead, manganese, mercury, nickel, and selenium.
3. All types of solid chemical waste.

Statutory procedures for waste and effluent management take precedence over any other methods. However, for operations in which compliance with statutory regulations is exempt or inapplicable because of scale or other circumstances, the following suggestions may be helpful.

Gases

Noxious gases and vapors from volatile compounds are best dealt with at the point of generation by "scrubbing" the effluent gas. The gas being swept from a reaction set-up is led through tubing to a (large!) trap to prevent suck-back and into a sintered glass gas dispersion tube immersed in the scrubbing fluid. A bleach container can be conveniently used as a vessel for the scrubbing fluid. The nature of the effluent determines which of four common fluids should be used: dilute sulfuric acid, dilute alkali or sodium carbonate solution, laundry bleach when an oxidizing scrubber is needed, and sodium thiosulfate solution or diluted alkaline sodium borohydride when a reducing scrubber is needed. Ice should be added if an exotherm is anticipated.

Larger scale operations may require the use of a pH meter or starch/iodide test paper to ensure that the scrubbing capacity is not being exceeded.

When the operation is complete, the contents of the scrubber can be poured down the laboratory sink with a large excess (10–100 volumes) of water. If the solution is a large volume of dilute acid or base, it should be neutralized before being poured down the sink.

Liquids

Every laboratory should be equipped with a waste solvent container in which *all* waste organic solvents and solutions are collected. The contents of these containers should be periodically transferred to properly labeled waste solvent drums and arrangements made for contracted disposal in a regulated and licensed incineration facility.**

Aqueous waste containing dissolved toxic organic material should be decomposed *in situ*, when feasible, by adding acid, base, oxidant, or reductant. Otherwise, the material should be concentrated to a minimum volume and added to the contents of a waste solvent drum.

Aqueous waste containing dissolved toxic inorganic material should be evaporated to dryness and the residue handled as a solid chemical waste.

Solids

Soluble organic solid waste can usually be transferred into a waste solvent drum, provided near-term incineration of the contents is assured.

Inorganic solid wastes, particularly those containing toxic metals and toxic metal compounds, used Raney nickel, manganese dioxide, etc. should be placed in glass bottles or lined fiber drums, sealed, properly labeled, and arrangements made for disposal in a secure landfill.** Used mercury is particularly pernicious and small amounts should first be amalgamated with zinc or combined with excess sulfur to solidify the material.

Other types of solid laboratory waste including used silica gel and charcoal should also be packed, labeled, and sent for disposal in a secure landfill.

Special Note

Since local ordinances may vary widely from one locale to another, one should always check with appropriate authorities. Also, professional disposal services differ in their requirements for segregating and packaging waste.

**If arrangements for incineration of waste solvent and disposal of solid chemical waste by licensed contract disposal services are not in place, a list of providers of such services should be available from a state or local office of environmental protection.

PREFACE

The publication of Volume 82 of *Organic Syntheses* represents a major milestone for the organization – Prof. Jeremiah P. Freeman elected to retire as Secretary to the Board in 2004 after more than a generation of service. Under his leadership *Organic Syntheses* continued to grow in stature and became one of the most outstanding publications in the field of chemistry. It is difficult for anyone in the area to think of *Organic Syntheses* without linking it to Professor Freeman. All will miss the tremendous amount of knowledge, experience and dedication that he brought to this organization's success, and we wish him well in his future endeavors.

In response to his retirement and having to answer the question as to who will fill Professor Freeman's shoes and how should it be done, I am pleased as Volume Editor to welcome our new Editor-in-Chief, Professor Rick Danheiser, from MIT and our new Associate Editor, Professor Charles Zercher, from the University of New Hampshire as the new leaders of the *Organic Syntheses* Editorial Board. Rick will be serving as Editor-in-Chief for a five-year term, and this volume represents the first one brought to fruition under his and Charles's tenure. With more permanent editorial leadership, we hope to be in a position to rapidly respond to the numerous demands that the 21st century is already bringing to the organization.

This new volume of *Organic Syntheses* brings together an additional 28 checked and edited experimental procedures. Each relates to an important synthetic method or useful reagent, which the Editorial Board believes deserves highlighting in this publication. What is most impressive about the final array of procedures is that it is a microcosm of the depth and diversity of the chemical research that is part of modern organic chemistry.

Catalysis and enantioselection remain one of the principal areas of modern chemical research, and this is clearly reflected in the procedures in this volume. Included are a procedure for the **PREPARATION OF**

OPTICALLY ACTIVE (*R,R*)-HYDROXYBENZOIN FROM BENZOIN AND BENZYL employing the RuCl(*S*, *S*)-Tsdpen(*p*-cymene) catalyst under transfer hydrogenation conditions; a procedure for the **PREPARATION OF (*S,S*)-1,2-BIS (*tert*)-BUTYLMETHYL-PHOSPHINO) ETHANE AS A RHODIUM COMPLEX**; a procedure for the preparation of **AN EFFICIENT, HIGHLY DIASTEREO- AND ENANTIOSELECTIVE HETERO-DIELS-ALDER CATALYST**; a procedure for the preparation of **(2*S*)-(-)3-*exo*-(MORPHOLINO)-ISOBORNEOL**; a procedure for the **ASYMMETRIC ALCOHOLYSIS OF *MESO*-ANHYDRIDES MEDIATED BY ALKALOIDS** as applied to the cyclopentadiene-maleic anhydride Diels-Alder adduct employing chinchona alkaloids; a procedure detailing the **ASYMMETRIC REARRANGEMENT OF ALLYLIC TRICHLOROACETIMIDATES**; a procedure for the preparation of **(R_S)-(+)-METHYL-2-PROPANESULFINAMIDE**, an important chiral auxiliary; relative to the enantioselective synthesis of β-lactones a procedure for the **CATALYTIC ASYMMETRIC ACYL HALIDE-ALDEHYDE CYCLOCONDENSATION REACTION**; and a procedure for the **CATALYTIC REDUCTION OF AMIDES TO AMINES WITH HYDROSILANES USING A TRIBUTYLRUTHENIUM CARBONYL CLUSTER AS THE CATALYST**. An unusual enzyme resolution is noted in the procedure **LIPASE-CATALYZED RESOLUTION OF 4-TRIMETHYLSILYL-3-BUTYN-2-OL AND CONVERSION OF THE *R*-ENANTIOMER TO (*R*)-3-BUTYN-2-LY MESYLATE AND (*P*)-1-TRIBUTYLSTANNYL-1,2-BUTADIENE.**

Modern synthetic methods are captured in procedures for the **IRIDIUM-CATALYZED SYNTHESIS OF VINYL ETHERS FROM ALCOHOLS AND VINYL ACETATE; *ORTHO*-FORMYLATION OF PHENOLS: PREPARATION OF 3-BROMOSALICYLALDEHYDE; PREPARATION OF 1-METHOXY-2-(4-METHOXYPHENOXY)-BENZENE; CONVERSION OF ARYLALKYLKETONES INTO DICHLOROALKENES; A PRACTICAL AND SAFE PREPARATION OF 3,5-(TRIFLUOROMETHYL)ACETOPHENONE;** and a method for the **IRIDIUM-CATALYZED C-H BORYLATION OF ARENES AND HETEROARENES**.

The synthesis of important reagents and intermediates is noted in the following procedures: **PREPARATION OF 1,4:5,8-DIMETHANO-1,2,3,4,5,6,7,8-OCTAHYDRO-9,10-DIMETHOXYANTHRACENIUM HEXACHLOROANTIMONATE: A HIGHLY ROBUST RADICAL-CATION SALT; 2,2-DIETHOXY-1-ISOCYANO-**

ETHANE; PREPARATION OF HEXAKIS(4-BROMOPHENYL) BENZENE;1-(*tert*-BUTYLIMINOMETHYL)-1,3-DIMETHYL-UREA HYDROCHLORIDE; D-RIBONOLACTONE AND 2,3-ISOPROPYLIDENE-(D-RIBONOLACTONE); PREPARATION OF 4-ACETYLAMINO-1,1,6,6-TETRAMETHYLPIPERIDINE-1-OXOAMMONIUM TETRA FLUOROBORATE; 1,4-DIOXENE; (*R*)-(+)-3,4-DIMETHYLCYCLOHEX-2-EN-1-ONE; PREPARATION OF (TRIPHENYLPHOSPHORANYLIDENE)KETENE; and **PREPARATION OF [1-METHOXYMETHYLCARBAMOYL)ETHYL] PHOSPHONIC ACID BIS(2,2,2-TRIFLUOROETHYL) ESTER.**

As my tenure on the Editorial Board is coming to an end, I would like to express my deepest appreciation to Professor Freeman for all that he has brought to us throughout his career, a special thanks to my fellow Editorial Board members, and a wish for ***Calm Seas and a Prosperous Voyage*** to Professors Danheiser and Zercher in the journey that lies ahead.

EDWARD J. J. GRABOWSKI
Volume Editor

James Cason
August 30, 1912–November 3, 2003

James Cason, Emeritus Professor of Chemistry at the University of California, Berkeley, died November 3, 2003. He was born August 30, 1912, in Murfreesboro, Tennessee, and attended Vanderbilt University, where he obtained an A.B. Degree in 1934. He earned a M.S. Degree from the University of California at Berkeley (1935) and Ph.D. from Yale (1938), both in organic chemistry. He was awarded a postdoctoral fellowship to Harvard and worked with the National Defense Research Committee during World War II. He taught at DePauw University (1940–41) and at Vanderbilt University (1941–45), before joining the faculty at the University of California, Berkeley, in 1945.

From its early beginnings until the end of World War II, Berkeley had been characterized by exceptional strength in physical chemistry, mainly through the influence of Gilbert Newton Lewis, who passed away in 1944. In 1945, Dean Wendell Latimer recognized the need to offer a more balanced program and set about to recruit a cadre of organic chemists and subsequently hired Cason, William Dauben, and Henry Rapoport in 1945–46.

For almost four decades, Cason taught organic chemistry at Berkeley and served as Dean of the College of Chemistry in 1955–56. He authored four college textbooks on organic chemistry and published more than a hundred articles in major scientific journals. He served on the *Organic Syntheses* active board from 1951–1959 and edited Volume

37, which was published in 1957. His last publication was an autobiography, published three years before his death ("Things Remembered," Rutledge Books, Inc., Danbury, CT, 2000).

Cason retired from the University of California in 1983 and for the last twenty years of his life, he and his wife Rebecca split their time between their home in the Berkeley hills and their old-growth redwood property, named "Camelot," near Garberville, California. For a number of years, the Casons operated a profitable 75-acre almond orchard in the Central Valley of California. Rebecca Marsden Cason passed away in 2004.

CLAYTON H. HEATHCOCK

Kenji Koga
February 11, 1938–July 25, 2004

Kenji Koga, the fourth Editor from Japan elected to the Board of Editors of *Organic Syntheses* (1995–1998), died of cancer on July 25, 2004, in Tokyo, Japan. He was 66 years old.

Kenji Koga was born in Nagoya, Japan and grew up in Miyazaki and Fukuoka. He received his undergraduate and graduate education at the Faculty of Pharmaceutical Sciences, the University of Tokyo, where he received his bachelor's degree in 1960 and his Ph.D. in 1967, both under the direction of Shun-ichi Yamada. He continued his research in the laboratory of Yamada as an assistant professor (1965–1968) and then as an associate professor (1968–1976). He spent the period from 1971 to 1973 at the University of California at Los Angeles as a post-doctoral fellow in the laboratory of Donald J. Cram. In 1976, Koga was promoted to full professor as the successor to Yamada, and dedicated his efforts to research and education in pharmaceutical organic chemistry. He became an emeritus professor at the University of Tokyo in 1998, but continued his academic carrier at Nara Institute of Science and Technology (1993–2003) and at Waseda University (2003–2004).

Koga's early research focused on the studies of optically active amino acids as chiral sources for asymmetric synthesis. In particular, stereoselective hydride reduction and stereocontrolled deamination of amino acid esters were significant and versatile methods investigated by Koga for manipulating amino acid stereocenters in asymmetric synthesis. By application of these methods, a number of pharmaceutically useful chiral compounds have been synthesized from optically active amino acids, beginning with chloramphenicol in his B.S. thesis and culminating in polycyclic natural products such as maritidine, galanthamine, podorhizon, steganacin, verrucarinolactone, bourbonene, spatol and carbapenam. Further development in the chemistry of amino acids led to their use as chiral auxiliaries, particularly in asymmetric alkylation and conjugate addition at the α- or β-carbon of carbonyl compounds, exploiting chiral enamines or enimines derived from amino acid esters.

In addition to the asymmetric reactions and synthesis involving amino acids, Koga also made a significant contribution to host-guest chemistry. The use of cyclophanes for capturing organic guests in water was investigated. By use of x-ray crystallography Koga successfully obtained the first direct evidence of the ability of cyclophenes to form inclusion complexes. Water-soluble cyclophanes have since been used extensively in host-guest chemistry. Another significant contribution to host-guest chemistry was the peptide synthesis via alternative intramolecular aminolysis by crown ethers functionalized with two adjacent SH groups. This accomplishment is regarded as an outstanding representative example of synthetic reactions within enzyme model systems.

Through his studies on diastereomeric asymmetric reactions via chiral enamines or enimines, conformational control by metal chelation was confirmed to be the key factor for both diastereoselective and enantioselective reactions. After examinations of metal-chelated enantioselective reactions, including Lewis acid-catalyzed asymmetric Diels-Alder reactions, Koga focused on the chemistry of chiral lithium enolates formed with chiral amine bases derived from optically active amino acids. Asymmetric reaction with chiral bases was initially developed for asymmetric deprotonation of cyclohexanone, and then extended to kinetic resolution by deprotonation of 2-substituted cyclohexanones, as well as regio- and diastereoselective deprotonation of 3-keto steroids. The mechanism of the asymmetric induction was investigated thoroughly by ^{6}Li- and ^{15}N-NMR spectroscopy, revealing the structure of the chelated complex that is active for efficient asymmetric induction. Based on these fundamental studies, highly selective asymmetric induction was achieved for alkylation and protonation. In spite of the inherent difficulties, these asymmetric reactions have more recently been achieved catalytically by an exquisite combination of chiral base and achiral ligand.

Koga's contributions were recognized with numerous honors and awards as he was a leading contributor to organic synthesis, particularly as a pioneer of asymmetric synthesis using chiral bases. In 1988, he received the Inoue Prize for Science, and in 1994 he was the recipient of the Pharmaceutical Society of Japan Award. In 1995, Koga shared the Japan Academy Prize with Shun-ichi Yamada on "Novel Synthetic Methods of Optically Active Compounds Based on the Transcription of the Chirality of L-Amino Acids" and was honored nationally and internationally for his work. He also showed extraordinary devotion to education within the pharmaceutical sciences in Japan, demonstrating the significance of fundamental research in pharmaceutical education. Professor Kenji Koga was an active and intellectual leader of organic chemistry. As a man of sincerity and passion for research and education, he fostered a number of disciples who have made significant contributions to such new fields as catalytic asymmetric C-C bond-forming reactions, host-guest recognition at the membrane surface, and selective genome-targeting molecules. This memorable man is survived by his wife Sumiko of Tokyo, his son Yuji and his grandchildren Tsuyoshi and Risa of Ibaraki, Japan. Kenji Koga is dearly missed by his family, former coworkers, and his colleagues. His name will be memorialized in the Yamada-Koga Prize of the Japan Research Foundation for Optically Active Compounds.

KAZUNORI ODASHIMA and HISASHI YAMAMOTO

Blaine C. McKusick
March 22, 1918–January 4, 2005

Blaine McKusick was born on March 22, 1918 to James Gillespie Blaine McKusick and Marjorie Chase McKusick in Minneapolis, MN. He went to the local public schools and then attended the University of Minnesota where he received a Bachelor degree in Chemical Engineering in 1940. He went on to the University of Illinois where he studied organic chemistry and obtained a Ph.D. degree under Professor Fuson in 1944. Subsequently he pursued postdoctoral research at Harvard University where he met his future wife, Marjorie Jane Kirk, who at the time was in medical school. They were married in 1952.

Blaine had a noteworthy career in hands-on chemistry. While at Illinois, his dissertation research involved a study of mustard gas derivatives as part of a war-related project. This was followed by a postdoctoral position at Harvard working on insect repellents, also connected with the war effort. In 1945 he joined the Central Research Department of the DuPont Company where he worked in the laboratory on many projects including high pressure polyethylene synthesis, the synthesis of gem-dithiols from hydrogen disulfide and aldehydes and ketones at high pressures, the synthesis and properties of tetracyanoethylene dyes, and pioneering research on finding uses for the van der Graaf electron accelerator in synthetic chemistry. During this part of his career, he applied for and was granted a Guggenheim Fellowship in 1950 to pursue a year of postdoctoral research on the structure of alkaloids at the

Swiss Federal Institute of Technology (ETH) in Zurich under Professor Vladimir Prelog.

After his laboratory tenure, Blaine had an outstanding management career in DuPont's Central Research Department. In the late 1950's, he became a Research Supervisor overseeing research primarily in organic and organo-metallic chemistry. In the early 1960's, he assumed the position of a Laboratory Director directing several groups working in the same areas.

Following his tenure in the Central Research Department, Blaine became a Director of Research in DuPont's Agricultural Chemicals Department and then went on to become a director of DuPont's Haskell Laboratory for Toxicological and Industrial Medicine. He retired from this position in 1982. Over his long and productive career with DuPont, McKusick authored 40 technical papers, received 20 U. S. patents, and was responsible for 3 National Research Council reports.

Blaine's professional activities extended far beyond his career with DuPont. Over many years he was very active in local (Wilmington, DE) professional organizations such as the ACS, and the Wilmington Organic Chemists Club. On the national and international levels, he was active in the AAAS, *Organic Reactions, Organic Syntheses*, IUPAC, and he served on several ACS committees. After his retirement, he was involved in the activities of the College of Marine Studies at the University of Delaware, and in 1992 was given a Medal of Distinction by the University.

McKusick served *Organic Syntheses* in several capacities over many years. In 1957 he was elected to the Board of Editors and was editor of Volume 43 in 1963. He also served on the Board of Directors for a long period ending in 1996, and was president of Organic Syntheses, Inc. from 1989 to 1992.

Blaine received numerous rewards for his achievements. In addition to the Guggenheim Fellowship and the Medal of Distinction mentioned above, he received the ACS Award in Chemical Health and Safety in 1986, and a DuPont Safety and Health Award in 1990. These awards reflect his active involvement in laboratory safety issues during his working career and long after.

Marjorie, a prominent pediatrician, and Blaine had three children – Kirk, a computer scientist residing in Berkeley, California; James, a professor of English at the Baltimore campus of the University of Maryland; and Kathleen McKusick, a computer specialist in biotechnology living in Redwood City, California. Kathleen's daughter, Marjorie Rose, is Blaine's only grandchild. Marjorie McKusick passed away in

1976. Blaine then married Virginia Walters, an artist and teacher, who died in 1997, and in 1999 he married Emily Morris. He is survived by his wife Emily, his three children, his granddaughter Marjorie Rose, a sister Laura Bell Berthold, and a brother Marshall.

Blaine was deeply involved in many local organizations not related to his employment. Among these were the local Unitarian Church, a Science Discussion Group, a Great Books Group which he led for many years, and the Delaware Adolescent Program, Inc. (D. A. P. I.), which his wife Marjorie had co-founded many years ago. Members of these organizations will sorely miss Blaine McKusick.

JOHN F. HARRIS

CONTENTS

O
1) cyclopentadiene
2) H_2, Pd/C
3) Br_2
O
OH
OH
1) $(MeO)_2SO_2$, KOH
2) $SbCl_5$
OMe
OMe
$SbCl_6^-$

Ts
C_6H_5
N
Ru
Cl
C_6H_5
N
H_2
O
C_6H_5
C_6H_5
OH
RuCl[(*S,S*)-Tsdpen](*p*-cymene)
HCOOH/N$(C_2H_5)_3$, DMF
40 °C, 48 hr
OH
C_6H_5
C_6H_5
OH

OEt
EtO
NH_2
$HCO_2C_3H_7$
reflux
OEt
EtO
H
N
H
O
PPh_3, CCl_4, NEt_3
CH_2Cl_2, reflux
OEt
EtO
N≡C

t-$BuPCl_2$
1) 2 MeMgBr; then BH_3
2) sec-BuLi, (+)-sparteine; then CuCl2
BH_3
BH_3
t-Bu
P
Me
P
Me
t-Bu
1) CF_3SO_3H; then KOH
2) $[Rh(nbd)_2]BF_4$
BF_4^-
Rh
t-Bu
P
Me
P
Me
t-Bu

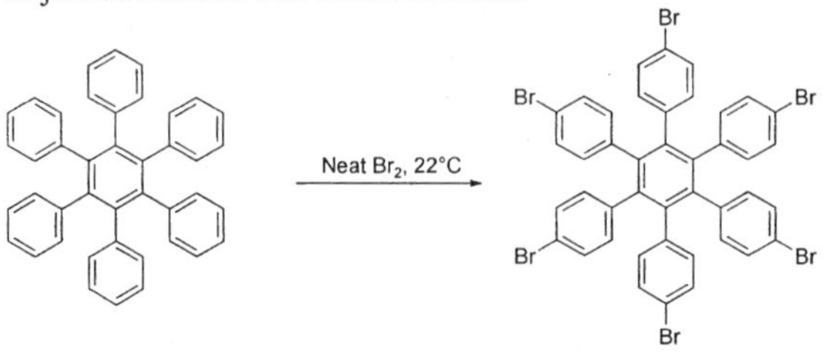

Neat Br_2, 22°C
Br

HPB HBB

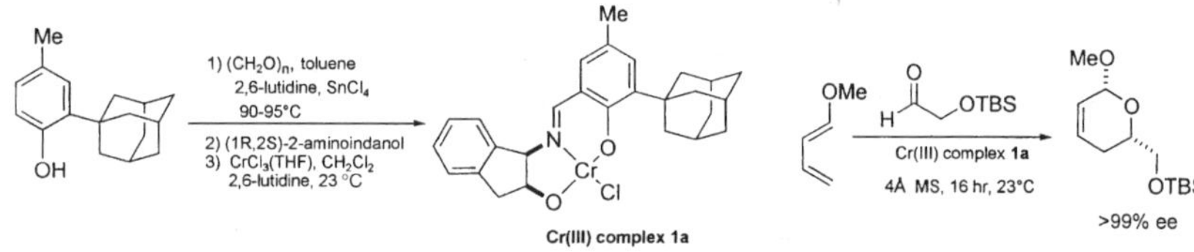

1) $(CH_2O)_n$, toluene 2,6-lutidine, $SnCl_4$ 90-95°C
2) (1R,2S)-2-aminoindanol
3) $CrCl_3(THF)$, CH_2Cl_2 2,6-lutidine, 23 °C
Cr(III) complex 1a
4Å MS, 16 hr, 23°C
>99% ee

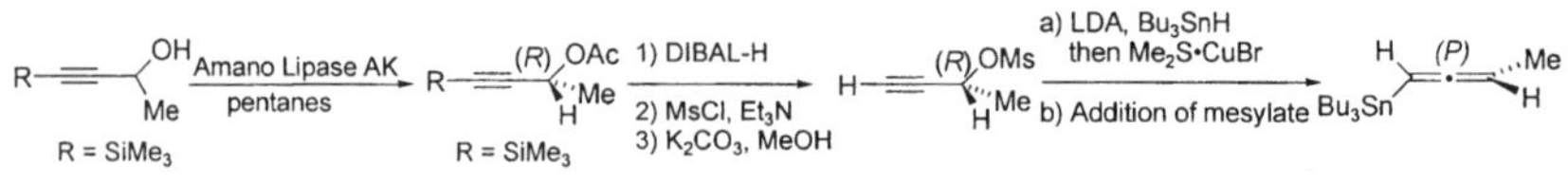

Amano Lipase AK
pentanes
R = $SiMe_3$
1) DIBAL-H
2) MsCl, Et_3N
3) K_2CO_3, MeOH
a) LDA, Bu_3SnH then Me_2S•CuBr
b) Addition of mesylate

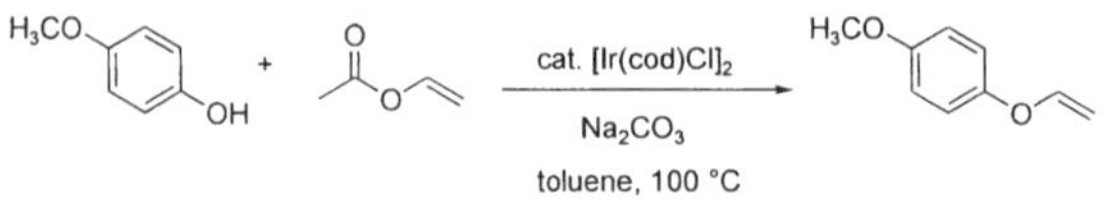

cat. $[Ir(cod)Cl]_2$
Na_2CO_3
toluene, 100 °C

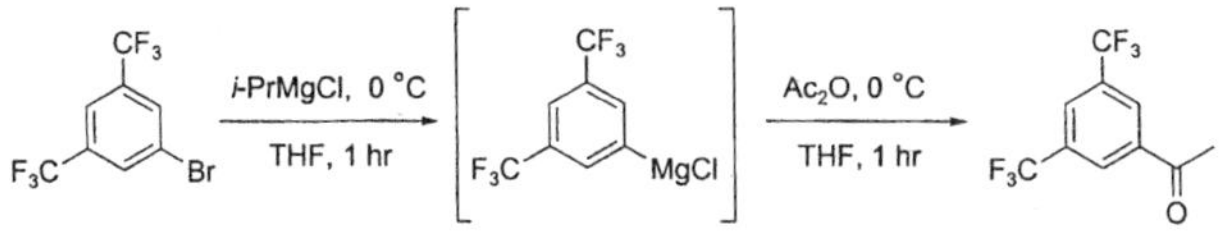
CF3
F3C
Br
i-PrMgCl, 0 °C
THF, 1 hr
CF3
F3C
MgCl
Ac2O, 0 °C
THF, 1 hr
CF3
F3C
O

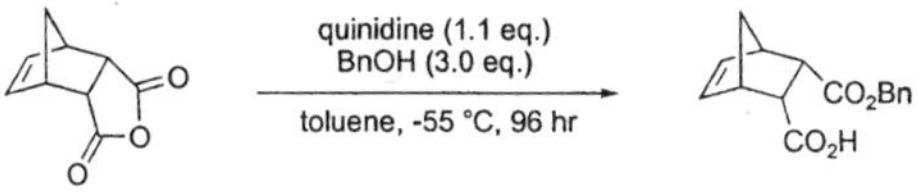
O
O
O
quinidine (1.1 eq.)
BnOH (3.0 eq.)
toluene, -55 °C, 96 hr
CO2Bn
CO2H

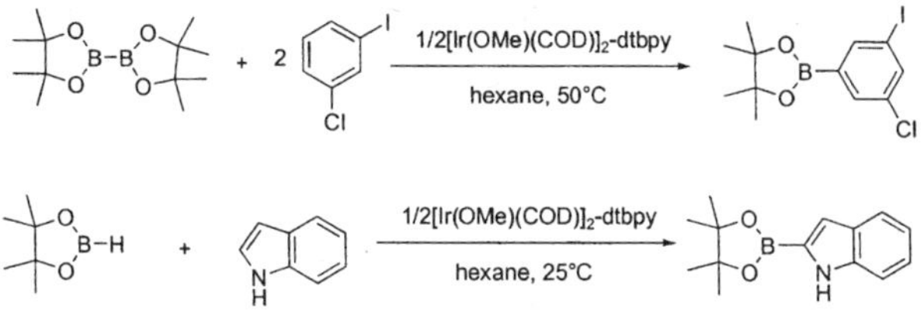
B−B
+ 2
I
Cl
1/2[Ir(OMe)(COD)]2-dtbpy
hexane, 50°C
B
I
Cl
B−H
+
N
H
1/2[Ir(OMe)(COD)]2-dtbpy
hexane, 25°C
B
N
H

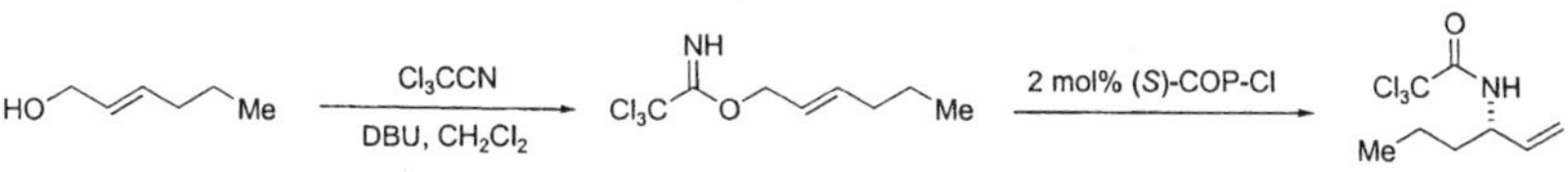
HO
Me
Cl3CCN
DBU, CH2Cl2
NH
Cl3C
O
Me
2 mol% (S)-COP-Cl
O
Cl3C
NH
Me

PREPARATION OF 1,4:5,8-DIMETHANO-1,2,3,4,5,6,7,8-OCTAHYDRO-9,10-DIMETHOXYANTHRACENIUM HEXACHLOROANTIMONATE ($4^{+\bullet}SbCl_6^-$): A HIGHLY ROBUST RADICAL-CATION SALT

[1,4:5,8-Dimethanoanthracene, 1,2,3,4,5,6,7,8-octahydro-9,10-dimethoxy-, radical ion(1+), hexachloroantimonate(1-)]

A. Ethanol, 0 °C → 1

B. 1 → EtOAC, Ethanol, Pd/C, H_2 → 2

C. 2 → Br_2, (1 equiv), $CHCl_3$ → 3

D. 3 → Ethanol. KOH, $(MeO)_2SO_2$ → 4

E. 4 + 3 $SbCl_5$ → CH_2Cl_2, -78 °C → $4^{\bullet+}$ $SbCl_6^-$ + $SbCl_3$

Submitted by Rajendra Rathore,[1] Carrie L. Burns, and Mihaela I. Deselnicu.
Checked by Scott E. Denmark and Tommy Bui.

1. Procedure

A. *1,4:5,8-Dimethano-1,4,4a,5,8,8a,9a,10a-octahydroanthracene-9,10-dione (**1**).* To a 500-mL conical flask is added *p*-benzoquinone (10.8 g, 100 mmol) (Note 1) and 100 mL of absolute ethanol. The mixture is stirred at room temperature for 10 min and then is cooled in an ice-salt bath (ca. –5°C, external) (Note 2). Freshly cracked cyclopentadiene (13.2 g, 200 mmol) (Note 3) is added slowly via syringe over 3-5 min and the resulting mixture is stirred in the ice-salt bath for 20 min and then for an additional 30 min at room temperature. A large amount of the adduct (**1**) precipitates during the course of the reaction (Note 4). The resulting slurry is cooled in an ice bath (ca. 0°C, external) and the precipitate is filtered using a Büchner funnel and is washed with ice-cold ethanol to yield 20.9-21.6 g (87-90%) of cycloadduct **1** as a nearly colorless, crystalline solid (Note 5).

B. *1,4:5,8-Dimethanododecahydroanthracene-9,10-dione (**2**).* A solution of diketone **1** (19.2 g, 80 mmol) in a mixture of 50 mL of ethanol and 250 mL of ethyl acetate is combined with 10% Pd-C (200 mg) in a 500-mL pressure bottle (Note 6). The bottle is pressured to 60 psi of hydrogen and shaken on a Parr apparatus for 3 hr (when the uptake of hydrogen ceases) (Note 7). The mixture is then diluted with 200 mL of dichloromethane to dissolve the precipitate and the solution is filtered *in vacuo* through a short pad of Celite (100 g) in a sintered glass funnel. The Celite pad is washed with dichloromethane (3 x 150 mL) and the colorless solution is concentrated on a rotary evaporator to afford diketone **2** (18.8 g, 96%) as a white solid (Note 8).

C. *1,4:5,8-Dimethano-1,2,3,4,5,6,7,8-octahydroanthracene-9,10-diol (**3**).* In a 500-mL, three-necked, round-bottomed flask equipped with a 125-mL pressure-equalizing addition funnel and nitrogen inlet and outlet adapters, is placed diketone **2** (18.3 g, 75 mmol) and 75 mL of anhydrous chloroform (Note 9). A solution of Br_2 (12.0 g, 75 mmol) in 75 mL of chloroform was placed in the dropping funnel and was added dropwise over 0.5 h under a slow stream of nitrogen at room temperature (Note 10). Stirring the reaction mixture while purging the flask with nitrogen is continued for another hour to remove the gaseous HBr. The resulting suspension is cooled in an ice-salt bath (approx. -5°C) and is filtered *in vacuo* using a Büchner funnel. The solid is washed with ice-cold chloroform (2 x 25 mL) and is suction dried to afford 17.4 g (96%) of hydroquinone **3** as a colorless powder, which is used in the next step without further purification.

D. *1,4:5,8-Dimethano-1,2,3,4,5,6,7,8-octahydro-9,10-dimethoxy-anthracene (**4**).* A solution of hydroquinone **3** (16.9 g, 70 mmol) in 175 mL of absolute ethanol is placed in a three-necked, 500-mL, round-bottomed flask equipped with a pressure-equalizing addition funnel and nitrogen inlet and outlet adapters. To this mixture is added a solution of KOH (8.7 g, 155 mmol) in 8.7 mL of water followed by a solution of sodium hydrosulfite (2 g) in 10 mL of water under an inert atmosphere. Dimethyl sulfate (19.6 g, 155 mmol) is placed in the addition funnel and is then added dropwise over 30 min. After the addition is complete, the addition funnel is replaced with a condenser and the mixture is heated to reflux in a 120°C oil bath for 3 h and then is cooled to room temperature. A second portion of KOH (2.0 g, 35 mmol, 0.5 equiv) is added to the flask followed by the dropwise addition of a second portion of dimethyl sulfate (4.4 g, 35 mmol, 0.5 equiv) through a 25-mL addition funnel. The addition funnel is replaced with a condenser and the mixture is heated for one hour and then is allowed to cool to room temperature. The reaction mixture is diluted with 250 mL of water and then is extracted with dichloromethane (3 x 100 mL). The combined dichloromethane extracts are successively washed with 10% aqueous NaOH solution (50 mL), water (100 mL), brine (50 mL), and then are dried over anhydrous magnesium sulfate. Removal of solvent *in vacuo* affords a crude material that is purified by filtration through a short column of silica gel (200 g) as follows. A solution of the crude product in a 25:1 mixture of hexanes and ethyl acetate (90 mL) is eluted through the silica gel column followed by an additional 900 mL of hexane/ethyl acetate, 25:1. The product-containing fractions are concentrated on a rotary evaporator to afford dimethyl ether **4** (15.3 g, 81%) as a cream colored solid (Notes 11 and 12).

E. *Preparation of* $\mathbf{4^{+\bullet}SbCl_6^-}$. A solution of hydroquinone ether **4** (2.7 g, 10 mmol) in 20 mL of anhydrous dichloromethane (Note 13) is placed in a 100-mL, three-necked, round-bottomed flask equipped with a 50-mL pressure-equalizing addition funnel and nitrogen inlet and outlet adapters. The addition funnel is charged with 15 mL of a 1.0 M solution of $SbCl_5$ in dichloromethane (15 mmol) (Note 14) and the flask is cooled in a dry ice-acetone bath (approximately –75 °C, external). The $SbCl_5$ solution is slowly added (3-5 min) under a flow of nitrogen. The reaction mixture immediately turns red and a large amount of dark-red material precipitates. The resulting mixture is warmed to 0°C over 5-10 min, and anhydrous diethyl ether (50 mL) is added to precipitate the dissolved $\mathbf{4}^{+\bullet}SbCl_6^-$ salt. The dark-red microcrystalline precipitate is suction-filtered using a medium-porosity sintered-glass funnel under a blanket of dry nitrogen and is washed with dry diethyl ether (2 x 20 mL) (Note 15). The salt is dried *in vacuo* at room temperature to afford $\mathbf{4}^{+\bullet}SbCl_6^-$ (5.6 g, 93%) as a red, crystalline solid (Notes 16 and 17).

2. Notes

1. *p*-Benzoquinone, 98% was purchased from Aldrich Chemical Co. and was used without further purification.

2. *p*-Benzoquinone, generally, precipitates upon cooling and it dissolves eventually as the cycloaddition reaction proceeds.

3. Cyclopentadiene is prepared by heating commercial dicyclopentadiene (available from Aldrich Chemical Company, Inc.) at 160°C in a distillation apparatus. Cyclopentadiene distills smoothly at 39–45°C. For a detailed procedure, see: Moffett, R. B. *Org. Synth., Coll. Vol. IV*, **1963**, 238.

4. This adduct can be directly hydrogenated (using a Parr hydrogenation apparatus) after dissolving the precipitate by adding 200-mL ethyl acetate and 10% Pd-C (200 mg, Aldrich Chemical Co.); however, a longer hydrogenation time is required. Furthermore, a less pure product that required recrystallization was afforded. A simple filtration of the adduct **1** followed by its hydrogenation (as described in the procedure above) yields a colorless solid of high purity.

5. The spectral data for 1,4:5,8-dimethano-1,4,4a,5,8,8a,9a,10a-octahydroanthra-cene-9,10-dione (**1**) are as follows: mp 157-158°C; ^1H NMR ($CDCl_3$) δ: 1.29 (br d, 2 H, J = 8.6 Hz), 1.45 (dt, 2 H, J = 8.6, 1.72 Hz), 2.87 (br, 4 H), 3.35 (br, 4 H), 6.18 (t, 4 H, J = 1.7 Hz); ^{13}C NMR ($CDCl_3$) δ: 48.2, 49.5, 53.2, 136.4, 212.7.

6. A standard hydrogenation bottle purchased from ACE glass Inc. was used.

7. The hydrogenation can also be carried out at atmospheric pressure as well.

8. The spectral data for 1,4:5,8-dimethanododecahydroanthracene-9,10-dione (**2**) are as follows: mp 226-229°C (decomp.); ^1H NMR ($CDCl_3$) δ: 1.25-1.30 (m, 4 H), 1.38-1.45 (m, 4 H), 1.38-1.58, (m, 4 H), 2.80 (br, 4 H), 2.86 (br, 4 H); ^{13}C NMR ($CDCl_3$) δ: 25.0, 39.2, 43.7, 53.6, 214.5.

9. Ethanol-free, anhydrous chloroform, which can be readily prepared according to the procedure from Perrin,[9] is recommended for this step.

Anhydrous dichloromethane can also be used as the solvent for the bromination.

10. The nitrogen outlet was connected to gas-wash bottle containing a 20% (w/w) solution of sodium hydroxide.

11. The spectral data for 1,4:5,8-dimethano-1,2,3,4,5,6,7,8-octahydro-9,10-dimethoxyanthracene (**4**) are as follows: mp 116-117°C (decomp.); ^{1}H NMR ($CDCl_3$) δ: 1.15-1.19 (m, 4 H), 1.42-1.44 (br d, 2 H, J = 8.55), 1.66-1.69 (m, 2 H), 1.87 (br d, 4 H, J = 7.33 Hz), 3.54 (m, 4 H), 3.83 (s, 6 H); ^{13}C NMR ($CDCl_3$) δ: 27.0, 40.6, 48.9, 61.3, 137.5, 143.5.

12. The submitters described recrystallization of **4** by dissolving 10 g in boiling hexanes (150 mL) and allowing the product to precipitate by standing at room temperature overnight.

13. Anhydrous dichloromethane was obtained according to the procedure of Perrin.[9]

14. A 1 M solution of antimony pentachloride in dichloromethane can be purchased from Aldrich Chemical Co. This solution can also be prepared by dissolving neat $SbCl_5$ in anhydrous dichloromethane.

15. A large inverted funnel, connected to a nitrogen outlet, positioned above the sintered glass funnel is generally sufficient for maintaining an inert atmosphere during filtration of the cation radical salt (see sketch of the apparatus below).

16. The spectral data of [$4^{+\cdot}$ $SbCl_6^-$]: UV-vis (dichloromethane), λ_{max}

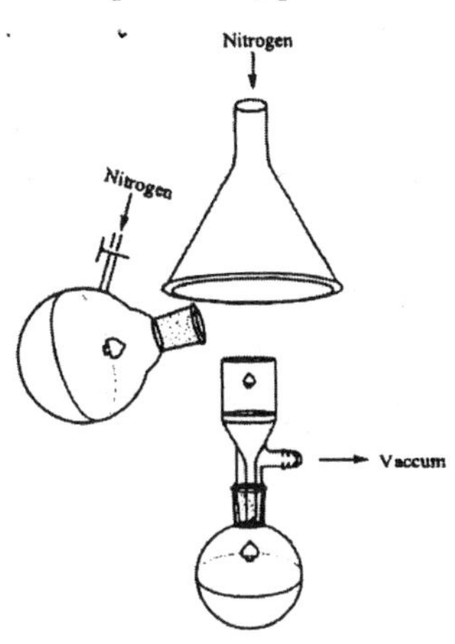

= 518 and 486 (shoulder) nm, extinction coefficient, ε_{518} = 7300 M^{-1} cm^{-1}. Anal. calcd. for $C_{18}H_{22}O_2$ $SbCl_6$: C, 35.74; H, 3.67; Cl, 35.17. Found: C, 35.64; H, 3.54; Cl, 34.16. The pure cation-radical salt ($4^{+\cdot}SbCl_6^-$) is >99%

can be stored in a glass bottle at ambient temperatures indefinitely. The purity of the $\mathbf{4}^{+\bullet}SbCl_6^-$ salt can be confirmed by iodimetric titration as follows. Thus, a solution of $\mathbf{4}^{+\bullet}$ $SbCl_6^-$ (60.45 mg, 0.01 M) in anhydrous dichloromethane was added to a dichloromethane solution containing excess tetra-*n*-butylammonium iodide (1 mmol, 0.1 M) at 22°C, under an argon atmosphere, to afford a dark brown solution. The mixture was stirred for 5 min and was titrated (with rapid stirring) by a slow addition of a standard aqueous sodium thiosulfate solution (0.005 M) in the presence of a starch solution as an internal indicator. Based on the amount of thiosulfate solution consumed (58.6 mL), purity of the cation radical was determined to be ca. 98.5%. Moreover, the purity can also be estimated by utilizing the spectrophotometric data as follows. Thus, a known quantity of $\mathbf{4}^{+\bullet}SbCl_6^-$ was dissolved in anhydrous dichloromethane and the absorbance was monitored at 518 nm using a 1-cm quartz cuvette. For example, a solution of 3.0 mg of $\mathbf{4}^{+\bullet}SbCl_6^-$ in 30 mL anhydrous dichloromethane showed an absorbance of 1.20 at 518 nm (actual concentration = 1.64×10^{-4} M, 99.6% pure). The actual concentration of the $\mathbf{4}^{+\bullet}SbCl_6^-$ was calculated by using the equation: Absorbance at 518 nm/7300 = actual concentration.

17. The radical cation salt $\mathbf{4}^{+\bullet}SbCl_6^-$ was also characterized by ESR spectroscopy. A solution of the salt (2 mg, 3.3 μmol) in 30 mL of dichloromethane was prepared in a dry box in a flame-dried, 50-mL round-bottomed flask. The solution was filtered through filter paper to remove particulates. About 0.25-0.35 mL of this solution was placed in a dry ESR tube under argon. The tube was sealed with a septum and the spectrum was recorded. G = 2.0046; A_1 = 3.3 Gauss (6H), A_2 = 2.1 Gauss (2H), A_3 = 1.2 Gauss (2H).

Safety and Waste Disposal Information

All hazardous materials should be handled and disposed of in accordance with "Prudent Practices in the Laboratory"; National Academy Press; Washington, DC, 1995.

3. Discussion

The hydroquinone ether **4** is an electron-rich molecule (E^o_{ox} = 1.11 V vs. SCE) that is readily oxidized to a remarkably stable cation-radical salt, which can be stored as a crystalline solid for an indefinite period. Stable radical-cation salts are of fundamental importance to organic materials science because they constitute the smallest units that carry both a delocalized positive charge and an unpaired electron. Thus, radical-cation salts provide the basis for experimentation in conductivity and ferromagnetism, etc. Radical cations are important intermediates in a rich menu of organic transformations.[2] They are also useful as electron-transfer catalysts[3] in a variety of organic and organometallic transformations, as well as tools for elucidation of electron transfer mechanisms.

Removal of an electron from a neutral molecule leads to an "umpolung" of the normal reactivity character. The cation radical possesses greatly enhanced electrophilic reactivity (due to the cationic charge), as well as homolytic reactivity (due to the radical character). Radical cations can promote a variety of unimolecular and bimolecular reactions as illustrated by following examples.[2]

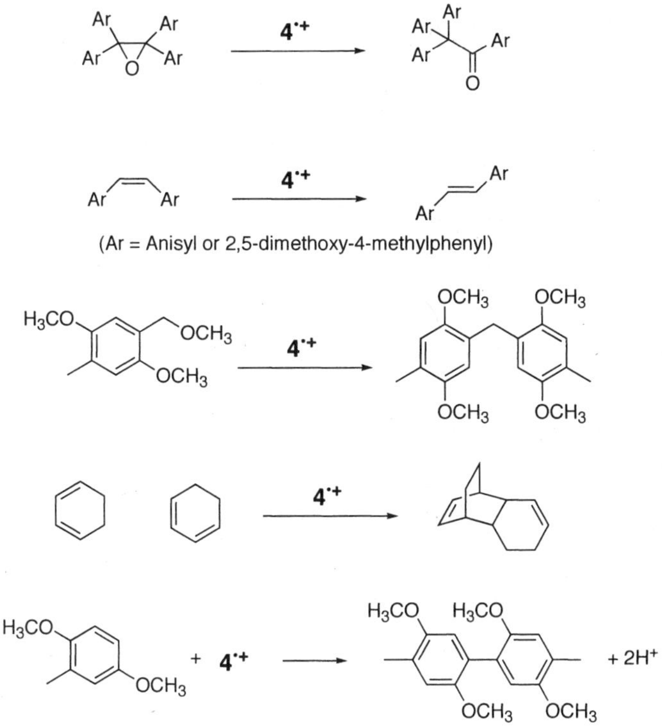

(Ar = Anisyl or 2,5-dimethoxy-4-methylphenyl)

In addition, the ready availability of $\mathbf{4}^{+\bullet}$ allows its utilization for the formation of cation radicals from a variety of organic donors for spectroscopic monitoring, as well as for the elucidation of electron transfer mechanisms.[3f,3g]

The method described here for the preparation of $4^{+\bullet}$ is essentially a detailed description of our recently published procedure.[4] To date, the only radical cation salts readily accessible to organic chemists are the heteroatom-centered *tris*(4-bromophenylaminium)[5] and thianthrenium[6] salts. It is envisioned that the ready availability of **4** and its cation radical $4^{+\bullet}$ will foster its use in a variety of organic and organometallic syntheses,[3,7] as well as in the exploration of the role of single electron transfer (SET) in various chemical processes.[2,8]

1. Department of Chemistry, Marquette University, P.O. Box 1881, Milwaukee, WI 53201-1881.
2. (a) Rathore, R.; Kochi J. K. *Adv. Phys. Org. Chem.* **2000**, *35*, 196. (b) Rathore, R.; Kochi, J. K. *Acta. Chem. Scand.* **1998**, *52*, 14.
3. (a) Chanon, M. *Bull. Soc. Chim. Fr.* **1985**, 209. (b) Bauld, N. L.; Bellville, D. J.; Harirchian, B.; Lorenz, K. T.; Pabon, P. A., Jr.; Raynolds, D. W.; Wirth, D. D.; Chiou, H. S.; Marsh, B. K. *Acc. Chem. Res.* **1987**, *20*, 371. (c) Rathore, R.; Burns, C. L.; Deselnicu, M. I. *Org. Lett.* **2001**, *3*, 2887. (d) Rathore, R.; Kochi, J. K. *J. Org. Chem.* **1995**, *60*, 7479. (e) Bard, A. J.; Ledwith, A.; Shine, H. J. *Adv. Phys. Org. Chem.* **1976**, *13*, 155 and references cited therein. (f) Rathore, R.; Burns, C. L. *J. Org. Chem.* **2003**, *68*, 4071. (g) Rathore, R.; Burns, C. L.; Deselnicu, M. I. *Org. Lett.* **2001**, *3*, 2887.
4. Rathore, R.; Kochi, J .K. *J. Org. Chem.* **1994**, *60*, 4399.
5. Bell, F. A.; Ledwith, A.; Sherrington, D. C. *J. Chem. Soc. C.* **1969**, 2719.
6. Bandish, B. K.; Shine, H. J. *J. Org. Chem.* **1977**, *42*, 561.
7. Connelly, N. G.; Geiger, W. E. *Chem. Rev.* **1996**, *96*, 877 and references therein.
8. Eberson, L. *Electron Transfer Reactions in Organic Chemistry;* Springer: New York, 1987 and references cited therein.
9. Perrin, D. D.; Armarego, W. L. F.; Perrin, D.R. *Purification of Laboratory Chemicals, 2nd Ed.*; Pergamon: New York, 1980.

Appendix
Chemical Abstracts Nomenclature (Registry Number)

p-Benzoquinone: 2,5-Cyclohexadiene-1,4-dione; (106-51-4)

Cyclopentadiene: 1,3-Cyclopentadiene; (542-92-7)

1,4:5,8-Dimethano-1,4,4a,5,8,8a,9a,10a-octahydroanthracene-9,10-dione:

1,4:5,8-Dimethanoanthracene-9,10-dione, 1,4,4a,5,8,8a,9a,10a-octahydro-, (1*R*,4*S*,4a*R*,5*S*,8*R*,8a*S*,9a*S*,10a*R*)-; (78548-82-0)

1,4:5,8-Dimethanododecahydroanthracene-9,10-dione: 1,4:5,8-Dimethanoanthracene-9,10-dione, dodecahydro-, (1α,4α,4aα,5β,8β,8aβ,9aα,10aβ)-; (2065-48-7)

Bromine; (7726-95-6)

1,4:5,8-Dimethano-1,2,3,4,5,6,7,8-octahydroanthracene-9,10-diol: 1,4:5,8-Dimethanoanthracene-9,10-diol, 1,2,3,4,5,6,7,8-octahydro-, (1α,4α,5β,8β)-; (130778-68-6)

Potassium hydroxide; (1310-58-3)

Sodium hydrogen sulfite: Sulfurous acid, monosodium salt: (7631-90-5)

Dimethyl sulfate: Sulfuric acid, dimethyl ester; (77-78-1)

1,4:5,8-Dimethano-1,2,3,4,5,6,7,8-octahydro-9,10-dimethoxyanthracene: 1,4:5,8-Dimethanoanthracene, 1,2,3,4,5,6,7,8-octahydro-9,10-dimethoxy-, (1*R*,4*S*,5*S*,8*R*)-; (322733-47-1)

Antimony pentachloride; Antimony chloride; (7647-18-9)

PREPARATION OF OPTICALLY ACTIVE (*R*,*R*)-HYDROBENZOIN FROM BENZOIN OR BENZIL

[1,2-Ethanediol, 1,2-diphenyl-, (1*R*,2*R*)-]

A.

RuCl[(*S,S*)-Tsdpen[(*p*-cymene)

$HCOOH/N(C_2H_5)_3$, DMF

40 °C, 48 hr

B.

RuCl[(*S,S*)-Tsdpen[(*p*-cymene)

$HCOOH/N(C_2H_5)_3$

40 °C, 24 hr

Submitted by Takao Ikariya,[1] Shohei Hashiguchi,[2] Kunihiko Murata,[3] and Ryoji Noyori.[4]

Checked by Peter Wipf and David Amantini.

1. Procedure

A. *A 100-g scale synthesis from rac-Benzoin:* A 1-L four-necked, round-bottomed flask equipped with a mechanical stirrer, a reflux condenser bearing an inert gas inlet tube, a thermometer and a dropping funnel is charged with 290 mL (2.08 mol) of triethylamine (Note 1). The triethylamine is cooled to 4°C in an ice bath and formic acid (97.0 mL, 2.57 mol) is added slowly (Note 2). To the mixture of formic acid and triethylamine at ambient temperature is added *rac*-benzoin (Note 3) (170 g, 0.801 mol), RuCl[1*S*,2*S*)-*N*-*p*-toluenesulfonyl-1,2-diphenylethanediamine]-(η^6-*p*-cymene) (Notes 4 and 5) (0.204 g, 0.321 mmol), and 80 mL of dry DMF (Notes 1 and 6). After the reaction mixture is stirred at 40°C for 48 hr, 300 mL of water is added at 0°C with stirring (Note 7). The pale pink precipitate is filtered through a Büchner funnel, washed with water (2 x 500 mL), and dried *in vacuo* to give a white solid in 97% yield (Note 8). The crude product is dissolved in 700 mL of hot methanol at 60 °C. A small amount of insoluble material is removed through filtration and the filtrate is cooled initially to room temperature and then to 0-5°C to provide white crystals. The crystalline product is isolated by filtration, washed with cooled (ice bath) 2-propanol (400 mL), and dried to provide 129.7 g of optically pure (*R*,*R*)-hydrobenzoin as white crystals (*dl* > 99%, 99.9% ee, Note 9). Concentration of the mother liquors and another recrystallization from methanol (100 mL) provides a second crop of the product, 19.1 g (*dl* > 99%, 99.9% ee, Note 9). The overall yield is 148.8 g (87%).

B. *A 10-g scale synthesis from benzil:* A 100-mL four-necked, round-bottomed flask equipped with a mechanical stirrer, a reflux condenser bearing an inert gas inlet tube, a thermometer, and a dropping funnel is charged with a mixture of formic acid (8.70 mL, 230 mmol) and triethylamine (19.0 mL, 136 mmol) in a similar manner to Procedure A. To this formic acid-triethylamine mixture at ambient temperature is added benzil (Note 3) (11.0 g, 52.3 mmol), and RuCl[(1*S*,2*S*)-*N*-*p*-toluenesulfonyl-1,2-diphenylethanediamine](η^6-*p*-cymene) (33.3 mg, 0.0524 mmol) (Note 6). After the reaction mixture is stirred at 40°C for 24 hr, 50 mL of water is added at 0°C with stirring. The pale pink precipitate is filtered through a Büchner funnel, washed with water (50 mL), and dried *in vacuo* to give a white solid in 95% yield. The crude product is dissolved in 50 mL of hot methanol at 60°C. The filtrate is cooled to room temperature and then cooled to -40°C to give white crystals. The crystalline product is isolated by filtration, washed with 10 mL of cooled (ice-bath) 2-propanol, and dried to provide 9.3 g of pure (*R*,*R*)-hydrobenzoin as white crystals (82%, 100% ee).

2. Notes

1. Triethylamine and formic acid were purchased from Kanto Chemical Company and used without further purification. The checkers used chemicals from Fisher and Fluka. An azeotropic mixture of triethylamine and formic acid is commercially available but could not be used for this reaction (see discussion section). Dry DMF was purchased from J. T. Baker.

2. The reaction of triethylamine with formic acid is exothermic and may proceed violently unless performed by controlled addition.

3. Benzoin and benzil were purchased from Kanto Chemical Company and used without further purification. The checkers used chemicals from TCI and Acros, respectively. Substituted benzoins were prepared by benzoin condensation of the corresponding ring-substituted benzaldehydes.[5]

4. The checkers used the following procedure for the preparation of (1*S*,2*S*)-*N*-*p*-toluenesulfonyl-1,2-diphenylethanediamine (TsDPEN): a dry

CH_2Cl_2 solution (10 mL) of *p*-toluenesulfonyl chloride (0.893 g, 4.69 mmol) was added dropwise over 5 h (syringe pump addition) to a mixture of (*S,S*)-DPEN (0.995 g, 4.69 mmol) and triethylamine (0.69 mL, 4.5 mmol) in dry CH_2Cl_2 (30 mL) at 0°C. After the reaction mixture was stirred at 0°C for 6 hr, the solution was washed with water (2 x 10 mL) and saturated NaCl solution (10 mL) and then dried with Na_2SO_4. The solvent was removed under reduced pressure to give 1.659 g of crude white solid product. Recrystallization from ethyl acetate (8 mL) gave 1.146 g (67% yield) of the desired product as white crystals.

5. Commercially available chiral Ru complexes from Kanto Chemical Company were used by the Submitters; however, the complexes were prepared by the Checkers following a modified literature procedure.[6] (*R*)-RuCl[(1*S*,2*S*)-*p*-TsNCH(C_6H_5)CH(C_6H_5)NH_2](η^6-*p*-cymene): A mixture of [$RuCl_2$(η^6-*p*-cymene)]$_2$[7] (0.651 g, 1.06 mmol), (*S,S*)-TsDPEN[8] (0.780 g, 2.13 mmol), and triethylamine (0.60 mL, 4.3 mmol) in 2-propanol (21 mL) was stirred at 80°C for 1 hr. The orange solution was concentrated and the resulting solid was collected by filtration, washed with a small amount of water and dried under reduced pressure to give the chiral Ru complex. After recrystallization from methanol (20 mL), 0.552 g of pure Ru (II) catalyst were collected as bright orange crystals. After two additional recrystallizations of the concentrated mother liquor from methanol (8.0 and 5.0 mL, respectively), an additional 0.327 g of pure catalyst were collected. The overall yield was 0.879 g (65%). mp > 100°C (dec.); IR (KBr) [cm^{-1}]: 3468, 3277, 3220 (H-N), 3062, 3029 (H–$C_{aromat.}$), 2961, 2872 (H–$C_{aliphat.}$); MS (EI): *m/z* (%) = 603 (63), 601 (100), 600 (66), 599 (58); ^1H NMR (300 MHz, $CDCl_3$, 25 °C, TMS) δ: 1.37 (m, 6H, CH(CH_3)$_2$), 2.21 (s, 3H, CH_3 in *p*-cymene), 2.35 (s, 3H, CH_3 in *p*-Ts), 3.12 (m, 1H, CH(CH_3)$_2$), 3.52 (m, 2H, N*H*H and *H*CNH_2), 3.73 (d, 1H, *J* = 10.5 Hz, *H*CN-*p*-Ts), 5.72 (m, 4H, $CH_{aromat.}$ in *p*-cymene), 6.04 (m, 1H, NH*H*), 6.34–7.05 (m, 14H, *p*-$CH_3C_6H_4$-SO_2NCH(C_6H_5)CH(C_6H_5)NH_2).

6. DMF was used to maintain the homogeneity of the reaction mixture, but it is not crucial for the catalysis to be efficient and practical. On small scale the addition of DMF is not needed. In fact, the reaction of benzil with a substrate to catalyst ratio (S/C) of 1,000 (4.7 M) in a mixture of HCOOH and N(C_2H_5)$_3$ containing the (*S,S*)-Ru catalyst (benzil:HCOOH:N(C_2H_5)$_3$ = 1:4.4:2.6) proceeded heterogeneously at the early stages of the reaction because of the low solubility of benzil. After about ten minutes, the reaction mixture changed to a completely homogenous solution, giving almost the same results as with DMF as solvent.

7. On large scale, the reaction flask should be connected to an Ar gas inlet to allow CO_2 to escape.

8. The diastereoselectivity of the product, *dl*:*meso* = 95.0:5.0, was determined by integration in the ^{1}H NMR (300 MHz, $CDCl_3$); (*R*,*R*)-hydrobenzoin, δ: 3.07 (s, 2H, OH), 4.86 (s, 2H, CH–OH), 7.25–7.40 (m, 10H, aromatic ring protons), *meso*-hydrobenzoin, δ: 2.33 (s, 2H, OH), 4.98 (s, 2H, CHOH), 7.30–7.45 (m, 10H, aromatic ring protons).

9. (*R*,*R*)-Hydrobenzoin: $[\alpha]_D^{25}$ + 91.6 (c 1.05, ethanol), (lit.[9a] $[\alpha]_D^{23}$, +95 (c 0.87 ethanol), 99% ee (*R*,*R*)). HPLC separation conditions, (column: CHIRALCEL OJ (4.6 mm i.d. x 250 mm), eluent: hexane/2-propanol = 90/10, flow rate: 1.0 mL/min, temp: 25 °C, detection UV 254 nm); retention time, (*S*,*S*)-hydrobenzoin, 14.2 min, (*R*,*R*)-hydrobenzoin, 16.5 min.

Safety and Waste Disposal Information

All hazardous materials should be handled and disposed of in accordance with "Prudent Practices in the Laboratory"; National Academy Press; Washington, DC, 1995.

3. Discussion

Optically active hydrobenzoins are useful building blocks for the stereoselective synthesis of various biologically active compounds, as well as chiral ligands and auxiliaries. The preparation of these chiral hydrobenzoins by Sharpless asymmetric dihydroxylation of trans-stilbene is one of the most convenient and well established methods.[9] Asymmetric reduction of readily available benzils or benzoins would be a promising and widely applicable approach; however, no practical reduction systems have been reported except for oxazaborolidine-catalyzed reductions of benzils with borane/methylsulfide.[10] Direct asymmetric hydrogenation of benzils or benzoins catalyzed by well-established Ru-BINAP complexes could potentially lead to optically active 1,2-diols. However, *meso*-isomers are obtained as the major products, because the substrate control of the hydroxy ketone intermediate favors *meso*-diol formation.[11] The procedure described herein provides a highly efficient method accessing the desired chiral hydrobenzoins in high enantiomeric excess using commercially available chiral (Ru(II) catalysts, RuCl(TsDPEN) (η^6-arene),[6,12] and easily handled reagents such as benzils or benzoins as substrates and a formic acid and triethylamine mixture as the hydrogen source.[13]

A mixture of formic acid and triethylamine is the best hydrogen donor for this reduction. In the absence of triethylamine, no conversion of benzoins or benzils was observed. The addition of triethylamine to the reaction mixture causes a significant increase in the conversion of the substrates. In the reaction of benzoin, a formic acid:triethylamine molar ratio of 3.2:2.6 to 3.2:4.4 gives the best catalyst performance in terms of both reactivity and stereoselectivity. The reduction of benzil requires a molar ratio of 4.4:2.6 to 4.4:4.4. The reaction with an azeotropic mixture of formic acid and triethylamine (5:2) gave no conversion under otherwise identical conditions as described in the Procedure.

The success of this asymmetric reduction of benzil or benzoin leading to the optically active hydrobenzoin with the formic acid and triethylamine mixture relies strongly on the nature of benzoin with a configurationally labile stereogenic center and the enantiomer discrimination ability of the chiral Ru complexes. Due to a sufficiently rapid stereomutation of benzoins under the basic reaction conditions, the dynamic kinetic resolution of benzoins allows the diastereo- and enantioselective synthesis of optically active hydrobenzoins.[13] Reduction of (*R*)-benzoin with the (*S*,*S*)-Ru catalyst in DMF under the same conditions gave (*R*,*R*)-hydrobenzoin quantitatively and in 100% ee, indicating that the (*S*,*S*)-Ru catalyst favors the reaction of (*R*)-benzoin.[13a] The rate of the reduction of (*R*)-benzoin with the (*S*,*S*)-Ru catalyst proceeds 55 times faster than the *S*-isomer. The slow-reacting *S*-isomer undergoes a rapid racemization.

Table 1. Asymmetic Reduction of Benzils with (*S*,*S*)-Ru Catalyst

R	S/C	temp, °C	time, h	yield, %	*dl:meso*	ee, %
H	1000	40	24	100	98.4:1.6	>99
p-CH_3	1000	40	48	67	96.7:3.3	>99
p-OCH_3[a]	200	35	48	75	94.4:5.6	>99
p-F	1000	40	24	100	94.2:5.8	>99

Conditions: (*S*,*S*)-Ru cat 0.005 mmol, ketone/HCOOH/$N(C_2H_5)_3$ = 1/4.4/2.6.
[a] ketone/HCOOH/$N(C_2H_5)_3$ = 4.4/4.4 in 1.2 M DMF.

Various benzil derivatives bearing substituents on aromatic rings can be reduced stereoselectively to the chiral hydrobenzoins in high ee's and in good yields (Table 1). The benzils with electron-donating substituents such as methyl or methoxy groups are reduced with excellent enantioselectivity but more slowly, while the reduction of *p*-fluorobenzil proceeded rapidly, as expected, giving a product with a high ee.[13]

The described, chiral Ru catalyst-promoted asymmetric transfer hydrogenation with a formic acid and triethylamine mixture is also applicable to the enantioselective reduction of acetophenone,[12] ring-

(*S,S*)-Ru cat
$HCOOH/N(C_2H_5)_3$, 28-40 °C, 16-20 hr

(S/C = 200)

R = H, X = H, >99% yield, 98% ee
R = H, X = CH_3, 96% yield, 97% ee
R = *m*-Cl, X = H, >99% yield, 97% ee
R = *p-Cl,* X = H, >99% yield, 95% ee
R = *m*-OCH_3, X = H, >99% yield, 96% ee
R = *p*-OCH_3, X = H, >99% yield, 97% ee
R = H, X = COC_6H_5, 99% yield, *dl/meso* = 94/6, 99% ee
R = H, X = CN, 100% yield, 98% ee (S/C = 1000)
R = H, X = N_3, 65% yield, 92% ee
R = H, X = NO_2, 90% yield, 98% ee (in 1.0 M DMF)
R = *p*-F, X = NO_2, 95% yield, 96% ee (in 1.0 M DMF)
R = *p*-CH_3, X = NO_2, 67% yield, 95% ee (in 1.0 M DMF)
R = H, X = Cl, 36% yield, 91% ee (S/C = 1000 in 1.0 M AcOEt)

(*S,S*)-Ru cat
$HCOOH/N(C_2H_5)_3$, 28°C, 48 hr

S/C = 200

R = CH_2, >99% yield, 99% ee
R = $(CH_2)_2$, 99% yield, 99% ee

(*S,S*)-Ru cat
$HCOOH/N(C_2H_5)_3$, 28°C, 40-65 hr

S/C = 200

X = S, 95% yield, 99% ee
X = SO_2, 95% yield, 98% ee

(*S,S*)-Ru cat: RuCl[(*S,S*)-Tsdpen](η^6-mesitylene) or RuCl[(*S,S*)-Tsdpen](*p*-cymene)

Figure 1. Examples of Chiral Ru Catalyzed Asymmetric Reduction of Ketones with $HCOOH/N(C_2H_5)_3$

substituted acetophenone derivatives,[12] α-substituted acetophenones,[14,15] acetylpyridine derivatives,[16] and functionalized ketones[17,18] leading to the corresponding optically active alcohols in excellent ee. These asymmetric reductions with the chiral Ru catalyst are characterized by a rapid, carbonyl group-selective transformation because of the coordinatively saturated nature of the diamine-based Ru hydride complexes.[6,17] The neighboring groups at the α-position of the carbonyl group do not interact with the metal center, leading to excellent reactivity and enantioselectivity. Some representative examples are listed in Figure 1.

1. Graduate School of Science and Engineering, Frontier Collaborative Research Center, Tokyo Institute of Technology, Meguro-ku, Tokyo 152-8552 Japan.
2. Takeda Chemical Industries, ltd., Yodogawa-ku, Osaka, Japan.
3. Kanto Chemical Corp. Inc., Central Research Laboratory, Soka, Saitama, Japan.
4. Department of Chemistry and Research Center for Materials Science, Nagoya University, Chikusa-ku, Nagoya, Japan.
5. Ide, W. S.; Buck, J. S. In "Organic Reactions", Adams, M.; Bachmann, W. E.; Blatt, A. H.; Fieser, L. F.; Johnson, J. R.; Snyder, H. R. ed. John Wiley & Sons, New York, **1948**, *4*, pp. 269-304.
6. Haack, K.-J.; Hashiguchi, S.; Fujii, A.; Ikariya, A.; Noyori, R. *Angew. Chem., Int. Ed. Engl.* **1997**, *36*, 285-288.
7. Bennett, M. W.; Huang, T.-N.,; Matheson, T. W.; Smith, A. K. *Inorg. Synth.* **1982**, *21*, 74-78.
8. Oda, T.; Irie, R.; Katuski, T.; Okawa, H. *Synlett* **1992**, 641-643.
9. (a) Wang, Z.-M.; Sharpless, K. B. *J. Org. Chem.* **1994**, *59*, 8302-8303. (b) Kolb, H. C.; VanNieuwenhze, M. S.; Sharpless, K. B. *Chem. Rev.* **1994**, *94*, 2483-2547.
10. Quallich, G. J.; Keavey, K. N.; Woodall, T. M.; *Tetrahedron Lett.* **1995**, *36*, 4729-4732.
11. Kitamura, M.; Ohkuma, T.; Inoue, S.; Sayo, N.; Kumobayashi, H.; Akutagawa, S.; Ohta, T.; Takaya, H.; Noyori, R. *J. Am. Chem. Soc.* **1988**, *110*, 629-631.
12. Hashiguchi, S.; Fujii, A.; Takehara, J.; Ikariya, T.; Noyori, R. *J. Am. Chem. Soc.* **1995**, *117*, 7562-7563.
13. (a) Murata, K.; Okano, K.; Miyagi, M.; Iwane, H.; Noyori, R.; Ikariya, T. *Org. Lett.* **1999**, *1*, 1119-1121. (b) Koike, T.; Murata, K.; Ikariya, T. *Org. Lett.* **2000**, *2*, 3833-3836.
14. (a) Hamada, T.; Torii, T.; Izawa, K.; Noyori, R.; Ikariya, T. *Org. Lett.* **2002**, *4*, 4373-4376. (b) Hamada, T.; Torii, T.; Izawa, K.; Ikariya, T. *Tetrahedron* **2004**, *60*, 7411-7417. (c) Hamada, T.; Torii, T.; Onishi, T.; Izawa, K.; Ikariya, T. *J. Org. Chem.* **2004**, *69*, 7391-7394.

15. Watanabe, M.; Murata, K.; Ikariya, T. *J. Org. Chem.* **2002**, *67*, 1712-1715.
16. Okano, K.; Murata, K.; Ikariya, T. *Tetrahedron Lett.* **2000**, *41*, 9277-9280.
17. Noyori, R.; Hashiguchi, S. *Acc. Chem. Res.* **1997**, *30*, 97-102.
18. (a) Palmer, M. J.; Wills, M. *Tetrahedron: Asymmetry* **1999**, *10*, 2045-2061. (b) Cross, D. J.; Kenny, J. A.; Houson, I.; Cambell, L.; Walsgrove, T.; Wills, M. *Tetrahedron: Asymmetry* **2001**, *12*, 1801-1806. (c) Mohar, B.; Valleix, A.; Desmurs, J.-R.; Felemez, M.; Wagner, A.; Mioskowski, C. *Chem. Commun.* **2001**, 2572-2573. (d) Eustache, F.; Dalko, P. I.; Cossy, J. *Org. Lett.* **2002**, *4*, 1263-1265.

Appendix
Chemical Abstracts Nomenclature (Registry Number)

Formic acid; (64-18-6)

Triethylamine: Ethanamine, *N,N*-diethyl-; (121-44-8)

rac-Benzoin: Ethanone, 2-hydroxy-1,2-diphenyl-; (19-53-9)

RuCl[(1*S*,2*S*)-*p*-TsNCH(C_6H_5)CH(C_6H_5)NH_2](η^6-*p*-cymene): Ruthenium, [*N*-[(1*S*,2*S*)-2-(amino-**κ***N*)-1,2-diphenylethyl]-4-methyl-benzenesulfonamidato-**κ***N*]chloro[(1,2,3,4,5,6-**η**)-1-methyl-4-(1-methylethyl)benzene]-; (192139-90-5)

(*R*,*R*)-Hydrobenzoin: 1,2-Ethanediol, 1,2-diphenyl-, (1*R*,2*R*)-; (52340-78-0)

2-Propanol; (67-63-0)

Benzil: Ethanedione, diphenyl-; (134-81-6)

2,2-DIETHOXY-1-ISOCYANOETHANE
(2,2-Diethoxyethyl isocyanide; Isocyanoacetaldehyde diethyl acetal)

$$\text{EtO-CH(OEt)-CH}_2\text{-NH}_2 \xrightarrow[\text{reflux}]{HCO_2C_3H_7} \text{EtO-CH(OEt)-CH}_2\text{-NH-CHO}$$

$$\text{EtO-CH(OEt)-CH}_2\text{-NH-CHO} \xrightarrow[CH_2Cl_2,\ \text{reflux}]{PPh_3,\ CCl_4,\ NEt_3} \text{EtO-CH(OEt)-CH}_2\text{-}N^{+}{\equiv}C^{-}$$

Submitted by Francesco Amato and Stefano Marcaccini.[1]
Checked by Raghuram S. Tangirala and Dennis P. Curran.

1. Procedure

CAUTION: All the operations must be conducted in an efficient hood because the isocyanide has an obnoxious odor.

A. *N-(2,2-Diethoxy)ethyl formamide.* A 100-mL round-bottomed flask equipped with a magnetic stir bar and fitted with a reflux condenser is charged with aminoacetaldehyde diethyl acetal (28.34 g, 213 mmol) (Note 1) and propyl formate (22.48 g, 255 mmol) (Note 2). The resulting clear solution is heated at reflux in an oil bath for 3 hr (Note 3). After cooling, the reaction mixture is transferred to a 250-mL round-bottomed flask and freed from the 1-propanol and the unreacted propyl formate by rotary evaporation. The residue is transferred to a distillation apparatus equipped with a 10 cm Vigreux column and a two-necked receiver, and distilled under reduced pressure. After a short forerun, the fraction boiling at 110-111°C (0.5 mmHg) is collected to give 29.3–29.5 g (86% yield) of *N*-(2,2-diethoxy)ethyl formamide (Notes 4, 5).

B. *2,2-Diethoxy-1-isocyanoethane.* A 500-mL one-necked flask equipped with a reflux condenser bearing a $CaCl_2$ trap at the upper end is charged with *N*-(2,2-diethoxy)ethyl formamide (24.20 g, 150 mmol), tetrachloromethane (24.61 g, 160 mmol) (Note 6), triphenylphosphine (44.59 g, 170 mmol) (Note 7), triethylamine (17.20 g, 150 mmol) (Note 8), and 150 mL of dichloromethane (Note 9). The clear mixture is heated at reflux in an oil bath, and a precipitate (triphenylphosphine oxide) begins to appear after 20-30 min. After 3.5 hr at reflux, the suspension is cooled to 5°C and filtered through a Büchner funnel under vacuum from a water aspirator. The collected solid is washed with 50 mL of diethyl ether. The filtrate and the washings are combined and evaporated to dryness, and the residue is stirred with a mixture of 100 mL of ethyl ether and 100 mL of pentanes (Note 10). The resulting suspension is allowed to stand overnight in the freezer (Note 11), and then filtered through a fritted funnel under vacuum with chilling of the collected filtrate in an ice-sodium chloride bath. The solid residue is washed with 60 mL of pentanes. The filtrate is concentrated on a rotary evaporator in a fume hood and the residue is transferred to a flask equipped with a short-path distillation head and a two-necked receiver. Distillation under reduced pressure gives 13.5–13.8 g (63-64%) of 2,2-diethoxy-1-isocyanoethane (Note 12), bp 60-61°C at 1 mmHg, as a colorless, vile-smelling liquid (Notes 13, 14).

2. Notes

1. Aminoacetaldehyde diethyl acetal was purchased from Aldrich and used as supplied.

2. Propyl formate (40 mL, purchased from Aldrich) is stirred with 20 mL aq. 5% $NaHCO_3$ for 2 min. The layers are separated, and the propyl formate layer is washed with 3 x 20 mL of distilled water then dried over magnesium sulfate. After filtration, the filtrate is distilled at atmospheric pressure and the fraction boiling at 80-81°C is collected for use in Step A.

3. The submitters used an electric shell for heating.

4. The distillation tends to bump, but the Vigreux column prevents overflow into the receiver.

5. IR spectrum (neat) 3304, 1666 cm^{-1}; ^{1}H NMR (300 MHz, $CDCl_3$), the formamide is an 6/1 ratio of amide rotamers in this solvent, major rotamer resonances δ: 1.29 (t, J = 7.1 Hz, 3 H, $CH_2C\underline{H_3}$), 3.45 (t, J = 5.4 Hz, 2 H, CH_2N), 3.50-3.77 (m, 4 H, $C\underline{H_2}CH_3$), 4.52 (t, J = 5.1 Hz, 1 H, O-CH-O), 5.81 (broad s, 1 H, NH), 8.21 (s, 1 H, NCHO); minor rotamer resonances δ: 3.31 (t, J = 6 Hz, 2 H, CH_2N), 4.45 (t, J = 6 Hz, 1 H, O-CH-O), 8.05 (d, J = 13.5 Hz, 1 H, NCHO); ^{13}C NMR (75 MHz, $CDCl_3$) major rotamer resonances δ: 14.9, 40.1, 62.4, 100.2, 161.3; minor rotamer resonances δ: 44.3, 63.0, 101.3, 165.0; LRMS (EI) *m/z* 117 (M – CH_3NO, 10%), 103 (100%), 91 (17%), 84 (55%), 75 (77%); HRMS (EI) *m/z* Calcd for $C_7H_{16}NO_2$ (M + H): 162.1130. Found: 162.1126.

6. Tetrachloromethane was purchased from Baker or Fisher and dried over molecular sieves before use.

7. Triphenylphosphine (99%) was purchased from Aldrich or Acros and used as supplied.

8. Reagent grade triethylamine (Fluka) was dried over calcium hydride pellets and distilled. The fraction boiling at 89°C was employed.

9. Dichloromethane was stored overnight on 4 Å molecular sieves (submitters) or distilled from calcium hydride (checkers) prior to use.

10. The lumps that formed were carefully broken with a spherical-ended glass rod.

11. The freezer temperature is about –15°C. If this operation is omitted, then additional solid that precipitates during the distillation makes this process difficult.

12. IR (neat) 2156 cm^{-1}; ^{1}H NMR (300 MHz, $CDCl_3$) δ: 1.17 (t, J = 7.0 Hz, 3 H, $CH_2C\underline{H}_3$), 3.43 (d, J = 5.4 Hz, 2 H, CH_2NC), 3.48-3.71 (m, 2 H, $C\underline{H}_2CH_3$), 4.64 (t, J = 5.4 Hz, 1 H, O-CH-O); ^{13}C NMR (75 MHz, $CDCl_3$) δ: 14.8, 44.3, 62.8, 99.1, 157.8.

13. The submitters obtained 71-75% yields. The submitters report that 2,2-diethoxy-1-isocyanoethane can be stored at –30°C under nitrogen for at least two years without appreciable decomposition. The checkers stored a sample at –20°C for three months, and the resulting liquid was still clear and exhibited a ^{1}H NMR spectrum identical to that recorded on the starting sample.

14. The checkers had the impression that this isonitrile smells fouler than phenyl isonitrile and related aryl isonitriles. Glassware can be freed from the isonitrile odor by rinsing with a 1:10 mixture of 37% hydrochloric acid/ethanol.

Waste Disposal Information

All hazardous materials should be handled and disposed of in accordance with "Prudent Practices in the Laboratory"; National Academy Press; Washington, DC, 1995.

3. Discusssion

This synthesis of 2,2-diethoxy-1-isocyanoethane is based on the dehydration of *N*-substituted formamides, which is the most important route to isocyanides.[2] The combination of triphenylphosphine, tetrachloromethane and triethylamine allows a smooth dehydration and a facile workup. The formylation of aminoacetaldehyde diethyl acetal employs propyl formate, because of the instability of acetals towards acidic reagents such as formic acid and formic-acetic anhydride that are usually employed in *N*-formylations.

Hartke[3] reported a synthesis of 2,2-diethoxy-1-isocyanoethane in which amino acetaldehyde acetal was transformed initially into the corresponding thioformamide. The thioformamide was then converted into the isocyanide by treatment with diphenylacetyl chloride/diisopropyl carbodiimide/triethylamine. The present method appears to be more convenient, because the experimental procedures are simpler, the yields are higher (62-68% overall) and the reagents are easily available and cheap.

This isocyanide has been employed as C–C–N–C unit in the synthesis of imidazoles, imidazo-imidazoles and aminoisoxazoles.[4]

1. Dipartimento di Chimica Organica "Ugo Schiff", Università di Firenze, via della Lastruccia 13, I-50019 Sesto Fiorentino (FI), Italy.
2. Ugi, I.; Dömling, A. *Angew, Chem. Int. Ed.* **2000**, *39*, 3168.
3. Hartke, K. *Chem. Ber.* **1966**, *99*, 3163.
4. (a) Bossio, R.; Marcaccini, S.; Pepino, R.; Polo, C.; Torroba, T. *Heterocycles* **1990**, *31*, 1287. (b) Bossio, R.; Marcaccini, S.; Pepino, R. *Liebigs Ann. Chem.* **1993**, 1289. (c) Bossio, R.; Marcaccini, S.; Pepino, R.; Torroba, T. *J. Org. Chem.* **1996**, *61*, 2202. (d) Buron, R.; El Kaïm, L.; Uslu, A. *Tetrahedron Lett.* **1997**, *46*, 8027.

Appendix
Chemical Abstracts Nomenclature (Registry Number)

Aminoacetaldehyde diethyl acetal: Ethanamine, 2,2-diethoxy-; (645-36-3)

Propyl formate: Formic acid, propyl ester; (110-74-7)

Tetrachloromethane: Methane, tetrachloro-; (56-23-5)

Triphenylphosphine: Phosphine, triphenyl-; (603-35-0)

Triphenylphosphine oxide: Phosphine oxide, triphenyl-; (791-28-6)

Triethylamine: Ethanamine, *N,N*-diethyl-; (121-44-8)

PREPARATION OF (*S,S*)-1,2-BIS(*tert*-BUTYLMETHYLPHOSPHINO)ETHANE ((*S,S*)-*t*-Bu-BISP*) AS A RHODIUM COMPLEX

[Rhodium(1+), [(2,3,5,6-η)-bicyclo[2.2.1]hepta-2,5-diene]bis(methyldiphenylphosphine)-, tetrafluoroborate(1-)]

A. *t*-$BuPCl_2$ — a) 2 MeMgBr; b) $BH_3 \cdot THF$ → BH_3, *t*-Bu–P(Me)Me

B. a) *sec*-BuLi, (-)-sparteine; b) $CuCl_2$

C. a) CF_3SO_3H; b) aq. KOH

D. $[Rh(nbd)_2]BF_4$ → Rh⊕, $BF_4^{\ominus}$

Submitted by Karen V. L. Crépy and Tsuneo Imamoto.[1]
Checked and substantially modified by Günter Seidel and Alois Fürstner.

1. Procedure

Caution! All reactions must be carried out in a well-ventilated hood.

A. *tert-Butyl(dimethyl)phosphine–Borane.* A 500-mL, three-necked flask equipped with a large football-shaped magnetic stirring bar, a thermometer, a three-way tap connected to an argon line, and a 250-mL pressure-equalizing dropping funnel fitted with a glass stopper is flame dried under vacuum and purged with argon (Note 1). The flask is charged with *tert*-butyldichlorophosphine (12.18 g, 77 mmol) (Note 2) and 120 mL of THF (Note 3) under argon and the resulting solution is cooled to −10°C using a cryostat. A solution of methylmagnesium bromide (3.0 M in Et_2O, 57 mL, 171 mmol) (Note 4) diluted with 100 mL of THF is added dropwise over 0.5 hr at such a rate as to maintain the internal temperature below 0°C, and the resulting heterogeneous mixture is stirred for 5 hr at room temperature. A solution of borane–THF complex (1.0 M in THF, 93 mL, 93 mmol) (Note 5) is then added dropwise over 20 min at −10°C and stirring is continued at room temperature overnight (18 hr). The mixture is cautiously poured into ice/water (120 mL) containing concentrated HCl (30 mL) (Note 6) and the resulting solution is vigorously stirred for 15 min. The layers are separated, the aqueous phase is extracted with ethyl acetate (2 x 100 mL), the combined organic layers are washed with brine (150 mL), dried over Na_2SO_4, filtered and evaporated under reduced pressure. The residue is recrystallized from 40 mL of hot hexane and dried under vacuum to give the *tert*-butyl(dimethyl)phosphine–borane adduct as a white solid (6.1 g, 61%) (Notes 7,8).

B. *(S,S)-1,2-Bis(boranato(tert-butyl)methylphosphino)ethane.* A 500-mL, three-necked flask equipped with a football-shaped magnetic stirring bar, a thermometer, a three-way tap connected to an argon line, and a 100-mL pressure-equalizing dropping funnel fitted with a glass stopper is flame-dried under vacuum and filled with argon. After cooling to ambient temperature, the flask is charged with (−)-sparteine (8.18 g, 35 mmol) and 40 mL of diethyl ether (Note 3) and the resulting solution is cooled to −78°C using a cryostat. *sec*-Butyllithium (1.3 M in cyclohexane, 26.8 mL, 35 mmol) is added dropwise at that temperature over 5 min and stirring is continued for 30 min before a solution of the *tert*-butyl(dimethyl)phosphine–borane adduct (4.21 g, 31.9 mmol) in 40 mL of diethyl ether is added dropwise over 30 min. The mixture is stirred at −78°C for 1 hr and −50°C for 4 hr. Copper(II) chloride (5.6 g, 41 mmol) (Note 9) is added with vigorous stirring and the resulting mixture is gradually warmed to room temperature overnight. Aqueous ammonia (25-28%, 40 mL) is added, the layers are separated, the bright blue aqueous phase is extracted with ethyl acetate (2 x 50 mL), the combined organic layers are successively washed with water (30 mL), aqueous HCl (3M, 2 x 30 mL), water (30 mL), and brine (30 mL) before being dried over Na_2SO_4 and evaporated under reduced pressure. The residue is recrystallized twice from 20 mL of toluene at 80°C (Note 10) to afford diastereomerically pure (*S,S*)-1,2-bis(boranato(*tert*-butyl)methylphosphino)ethane (2.0 g, 48%) as white needles (Note 11). The enantiomeric purity is checked by HPLC analysis using a chiral column (Note 12).

C. *(S,S)-1,2-Bis((tert-butyl)methylphosphino)ethane ((S,S)-t-Bu-BisP*)).* A 100-mL, two-necked flask equipped with a magnetic stirring bar, a three-way tap connected to an argon line, and a glass stopper is flame-dried under vacuum and filled with argon after reaching ambient temperature. The flask is charged with (*S,S*)-1,2-bis(boranato(*tert*-butyl)methylphosphino)ethane (1.14 g, 4.4 mmol) and 35 mL of toluene (Note 3) and the resulting mixture is cooled to 0°C. Trifluoromethanesulfonic acid (1.92 mL, 22 mmol) is added dropwise and the mixture is stirred for 15 min at 0°C and for 45 min at ambient temperature. The solvent is evaporated under reduced pressure before a solution of potassium hydroxide (2.44 g, 43 mmol) in 15 mL of freshly degassed ethanol/water (9:1 ratio) (Note 13) is added, and the resulting mixture is stirred at 50°C for 2 hr under argon. After reaching ambient temperature, the mixture is extracted four times with 20 mL of diethyl ether while keeping the positive pressure of argon. The combined organic phases are dried over Na_2SO_4 and passed through a short column of basic alumina (80 g) under argon (Note 14). The column is carefully rinsed with degassed diethyl ether (total volume 180 mL) and the combined eluents are evaporated to give a colorless pasty oil that solidifies on cooling to 0°C (980 mg, 95%). The free diphosphine is immediately subjected to complexation with a transition metal precursor.

D. *[Rh((S,S)-t-Bu-BisP*))(nbd)]BF₄.* A 250-mL, two-necked flask equipped with a magnetic stirring bar, a glass stopper, and a three-way tap connected to an argon line is flame-dried under vacuum and filled with argon. After reaching ambient temperature, the flask is charged with $[Rh(nbd)_2]BF_4$ (1.17 g, 3.1 mmol) (Note 15) and 100 mL of THF. A solution of (*S,S*)-1,2-bis((*tert*-butyl)methylphosphino)ethane (0.77 g, 3.3 mmol) in 40 mL of THF is added via syringe to this suspension and stirring is continued for 2 hr leading to the formation of an almost clear solution. This mixture is filtered under argon and the filtrate is evaporated. The orange powder is dissolved in hot THF until a clear solution is formed. This solution is slowly cooled to 0°C, leading to the precipitation of red cubes that are removed through filtration under argon. The filtrate is evaporated and the residue is recrystallized again by the same procedure. The combined crops of five recrystallizations are dried under vacuum to give the rhodium complex in analytically pure form (687 mg, 43%) (Notes 16, 17).

2. Notes

1. Argon (> 99.999%) was used by the checkers.

2. *tert*-Butyldichlorophosphine was purchased from Aldrich and used as received. *tert*-Butyldichlorophosphine is moderately air-sensitive and must be handled under argon.

3. THF, Et_2O, and toluene were freshly distilled from sodium/ benzophenone at atmospheric pressure under argon immediately prior to use.

4. The submitters used methylmagnesium bromide in THF (1.0M in THF) purchased from the Tokyo Kasei Company. The checkers purchased methylmagnesium bromide (3M in Et_2O) from Aldrich. It was transferred from the commercial bottle to the dropping funnel under argon *via* cannula.

5. Borane-THF complex was purchased from Kanto Chemical Company. The checkers purchased this reagent from Aldrich. Borane-THF is air-sensitive and must be handled under argon.

6. CAUTION: Excess methylmagnesium bromide and borane–THF complex react violently with water while liberating a large amount of gas. Slow addition under stirring is necessary. The use of a large Erlenmeyer flask is recommended.

7. The submitters reported a yield of 77-81% (19.2-20.4 g scale). The *tert*-butyl(dimethyl)phosphine–borane adduct prepared by this procedure shows the following physical and spectroscopic data: mp 164-165°C (hexane); R_f 0.44 (5:1 hexane/ethyl acetate); ^{1}H NMR (400 MHz, $CDCl_3$) δ: 0.44 [dq, 3 H, *J*(B,H) = 95 Hz, ^2J(P,H) = 15 Hz], 1.16 [d, 9 H, ^3J(P,H) = 14 Hz], 1.23 [d, 6 H, ^2J(P,H) = 9.9 Hz]; ^{13}C NMR (100 MHz, $CDCl_3$) δ: 7.3 [d, *J*(P,C) = 35.6 Hz], 24.7 [d, ^2J(P,C) = 2.2 Hz], 26.6 [d, *J*(P,C) = 35 Hz]; ^{11}B NMR (96 MHz, $CDCl_3$) δ: –39.9 [dq, *J*(B,H) = 95 Hz, *J*(P,B) = 63 Hz]; ^{31}P NMR (121 MHz, $CDCl_3$) δ: 20.9 [q, *J*(P,B) = 62 Hz]. IR (KBr): 2970, 2370, 1070, 945, 920 cm^{-1}.

8. The malodorous smell of the reaction/work-up glassware is removed by immersing all vessels for 1 hr in a mixture of domestic bleach and water (ca 1/10 ratio).

9. Finely powdered copper(II) chloride was dried for 2 hr at 130–140°C under vacuum prior to use.

10. The submitters report that 4-5 recrystallizations were necessary to obtain the desired (*S,S*)-*t*-Bu-BisP*–borane complex in pure form. It is important that the solid is carefully dried before proceeding to the next recrystallization.

11. The submitters reported yields of 60% (2.15 g scale) and 51-55% (5.3-5.6 g scale).

12. (*S,S*)-1,2-bis(boranato(*tert*-butyl)methylphosphino)ethane shows the following analytical and spectroscopic data: mp = 186-187.5°C; the submitters report a mp = 182-184°C. R_f 0.22 (5:1 hexane/ethyl acetate);

$[\alpha]^{28}{}_D$ −9.1 (c 1.21, chloroform); ee = 99.1% (250 mm Chiralcel OD-H, 0.5 mL/min; n-heptane/*i*-propanol = 9:1; RI detection; t = 15.3 min); IR (KBr): 2960, 2380, 2350, 1185, 1065, 765 cm^{-1}; ^1H NMR (400 MHz, $CDCl_3$) δ: 0.38 [q, 6 H, *J*(B,H) = 95 Hz], 1.17 [d, 18 H, ^3J(P,H) = 14 Hz], 1.21 [d, 6 H, ^2J(P,H) = 10 Hz], 1.60 (m, 2 H), 1.99 (m, 2 H). ^{13}C NMR (100 MHz, $CDCl_3$) δ: 5.6 [d, *J*(P,C) = 34 Hz], 15.9 [d, *J*(P,C) = 30 Hz], 25.1, 27.7 [d, *J*(P,C) = 34 Hz]; ^{11}B NMR (128 MHz, $CDCl_3$) δ: −41.1 [dq, *J*(P,B) = 63 Hz, *J*(B,H) = 94 Hz]; ^{31}P-NMR (162 MHz, $CDCl_3$) δ: 30.1. MS (EI) *m/z* (rel. intensity): 262 ([M^+], 4), 261 (25), 260 (15), 259 (27), 258 (15), 257 (20), 256 (9), 247 (100), 203 (24), 191 (21), 189 (55), 133 (17), 108 (5), 93 (6), 75 (5), 57 (42), 41 (41), 29 (17); HRMS (ESIpos) Calcd. For [M^+ + NH_4]): 280.26656. Found: 280.26674.

13. The checkers prepared the degassed solvents by two freeze/thaw cycles, whereas the submitters recommended the following procedure: The solvent is introduced in a two-necked flask fitted with a septum and a three-way tap connected to an argon balloon. It is placed in an ultra-sound bath and vacuum/argon cycles are applied three times.

14. Free phosphines are usually air-sensitive compounds and must be handled under argon. Therefore, all flasks must be evacuated and purged with argon three times before introducing any free phoshines. The alumina column must also be the subject of the same treatment (a thin pressure-equalizing dropping funnel is best used).

15. The checkers prepared the complex [Rh(nbd)$_2$]BF_4 in the following way: $AgBF_4$ (931 mg, 4.8 mmol) is added to a solution of [Rh(nbd)Cl]$_2$ (1.001 g, 2.17 mmol) and norbornadiene (1.8 mL, 16 mmol) in 30 mL of CH_2Cl_2 and the resulting mixture is stirred at ambient temperature for 4 hr under argon. The precipitate is filtered through a pad of Celite, the filtrate is evaporated, and the residue is triturated with Et_2O (50 mL). The solid material formed is filtered, rinsed with Et_2O (5 mL) and dried in vacuum to give [Rh(nbd)$_2$]BF_4 as a dark red, air sensitive, crystalline solid. Anal. Calcd. for $C_{14}H_{16}RhBF_4$: C, 44.96; H, 4.31; Found: C, 45.09; H, 4.35.

16. The submitters report a yield of 62% (2.24 g scale). The rhodium complex shows the following spectroscopic properties: ^1H NMR (400 MHz, $CDCl_3$) δ: 1.10 [d, 18 H, X-part of an $AA'X_9X'_9$ spin system, ^3J(P,H) = 14.5 Hz], 1.39 [d, 6 H, X-part of an $AA'X_3X'_3$ spin system (A = ^{31}P, X = ^1H, ^2J(P,H) = 8.3 Hz], 1.54 (m, 2 H), 1.83 (m, 2 H), 1.93 (m, 1 H), 2.04 (m, 1 H), 4.15 (m, 2 H), 5.73 (m, 4 H); ^{13}C NMR (75 MHz, $CDCl_3$, X-parts of an ABMX spin system (A, B = ^{31}P, M = ^{103}Rh, X = ^{13}C)) δ: 6.0 [*J*(Rh,C) = 1.3 Hz, *J*(P_A,P_B) = 21.5 Hz, *J*(P_A,C) = 20.8 Hz, *J*(P_B,C) = -0.6 Hz], 21.5 [*J*(Rh,C) = 2.8 Hz, *J*(P_A,P_B) = 21.4 Hz, *J*(P_A,C) = 27.8 Hz, *J*(P_B,C) = 10.9

Hz], 26.4 [$J(P_A,C) + J(P_B,C)$] = 4.2 Hz], 32.3 [$J(Rh,C)$ = 1.6 Hz, $J(P_A,P_B)$ = 21.4 Hz, $J(P_A,C)$ = 25.7 Hz, $J(P_B,C)$ = 0.8 Hz], 56.7 {$J(Rh,C)$ = 1.8 Hz, [$J(P_A,C) + J(P_B,C)$] = 3.4 Hz}, 72.3 {$J(Rh,C)$ = 3.9 Hz, [$J(P_A,C + J(P_B,C)$] = 5.7 Hz}, 85.0 {$J(Rh,C)$ = 6.2 Hz, [$J(P_A,C) + J(P_B,C)$] = 8.7 Hz}, 89.5 {$J(Rh,C)$ = 6.7 Hz, [$J(P_A,C) + J(P_B,C)$] = 9.2 Hz}; ^{11}B NMR (128 MHz, $CDCl_3$) δ: 0.0; ^{31}P NMR (121 MHz, $CDCl_3$): δ 62.0 [d, $J(Rh,P)$ = 152 Hz]. The structure was determined by X-ray crystallographic analysis.[2]

17. The rhodium complex is not readily oxidized on contact with air at room temperature, but it gradually decomposes on prolonged exposure to air. Therefore, storage under argon in a freezer is recommended. The checkers, however, found the complex to be unstable when kept in CD_3OD solution.

Waste Disposal Information

All hazardous materials should be handled and disposed of in accordance with "Prudent Practices in the Laboratory"; National Academy Press; Washington, DC, 1995.

3. Discussion

In contrast to optically active diphosphine ligands bearing a chiral-backbone, C_2-symmetric, P-stereogenic diphosphine ligands, such as (*S,S*)-1,2-bis(alkylmethylphosphino)ethane (also known as (*S,S*)-BisP*),[2] have been used only recently as ligands in enantioselective reactions. BisP* was found to be an excellent ligand for enantioselective hydrogenations of olefins or ketones. Its synthesis consists of only three steps starting from *tert*-butyldichlorophosphine. The asymmetric P-center is formed during the deprotonation step using *sec*-BuLi in the presence of (–)-sparteine.[3] Given the fact that the copper-promoted oxidative coupling[4] of the generated anion proceeds through a radical intermediate, the reaction leading to the (*S,S*) or (*R,R*)-isomers has almost the same rate as that leading to the *meso* compound.[5] Thus, (*S,S*)-BisP*–borane (>99% ee) is contaminated by a small amount of (*R,S*)-BisP*–borane which can be removed by recrystallization.

The (*R,R*)-enantionmer ((*R,R*)-*t*-Bu-BisP*)) may also be prepared via one of three protocols that we have developed.[6] The recommended method[6c] is reported in the scheme below:

It should be noted that tricoordinate phosphorus compounds in low oxidation states are usually air-sensitive, making their handling and storage difficult. Moreover, chiral phosphines bearing stereogenic phosphorus atoms are prone to racemization especially at higher temperatures[7] and therefore require additional stabilization. Unlike diaryl- and triarylphosphines, however, optically active trialkylphosphines hardly racemize even at elevated temperatures.[8] Temporary protection of the phosphine with BH_3 usually prevents all problems of this kind.[9] The phosphine–borane adducts are air-stable compounds which can be conveniently isolated, purified and stored. Owing to their remarkable inertness and resistance to a wide range of reactions conditions phosphine-borane complexes have emerged as indispensable intermediates for the preparation of P-stereogenic compounds.[10] Cleavage of the P–B bond releases the chiral phosphines with retention of configuration.[5] Complexation with a transition-metal prevents ready oxidation of the ligand.

1) (–)-sparteine /s-BuLi
2) $CuCl_2$
50%

$RuCl_3/KOH/K_2S_2O_8$
83%

1) n-BuLi
2) MeI
99%

(*R,R*)-*t*-Bu-BisP*-borane
98% ee (99% ee after recrystallization)

1. Department of Chemistry, Faculty of Science, Chiba University, Yayoi-cho, Inage-ku, Chiba 263-8522, Japan.
2. (a) Imamoto, T.; Watanabe, J.; Wada, Y.; Masuda, H.; Yamada, H.; Tsuruta, H.; Matsukawa, S.; Yamaguchi, K. *J. Am. Chem. Soc.* **1998,** *120,* 1635; (b) Gridnev, I. D.; Yamanoi, Y.; Higashi, N.; Tsuruta, H.; Yasutake, M.; Imamoto, T. *Adv. Synth. Catal.* **2001**, *343*, 118.
3. Muci, A. R.; Campos, K. R.; Evans, D. A. *J. Am. Chem. Soc.* **1995**, *117*, 9075.
4. Maryanoff, C. A.; Maryanoff, B. E.; Tang, R.; Mislow, K. *J. Am. Chem. Soc.* **1973**, *95*, 5839.
5. Imamoto, T.; Oshiki, T.; Onozawa, T.; Kusumoto, T.; Sato, K. *J. Am. Chem. Soc.* **1990**, *112*, 5244.
6. (a) Miura, T.; Yamada, H.; Kikuchi, S.; Imamoto, T. *J. Org. Chem.* **2000,** *65*, 1877; (b) Imamoto, T.; Kikuchi, S.; Miura, T.; Wada, Y. *Org. Lett.*

2001, *3*, 87; (c) Crépy, K. V. L.; Imamoto, T. *Tetrahedron Lett.* **2002**, *43*, 7735.
7. (a) Pietrusiewicz, K. M.; Zablocka, M. *Chem. Rev.* **1994**, *94*, 1375; (b) Valentine, D., Jr. In *Asymmetric Synthesis*; Morrison, J. D., Scott, J. W., Eds; Academic Press, Orlando, 1984; Vol. 4; Chapter 3.
8. Baechler, R. D.; Mislow, K. *J. Am. Chem. Soc.* **1970**, *92*, 3090.
9. (a) Ohff, M.; Holz, J.; Quirmbach, M.; Börner, A. *Synthesis* **1998**, 1391; (b) Carboni, B.; Monnier, L. *Tetrahedron* **1999**, *55*, 1197.
10. Imamoto, T. *Pure Appl. Chem.* **1993**, *65*, 655.

Appendix
Chemical Abstracts Nomenclature

tert-Butyldichlorophosphine: Phosphonous dichloride, (1,1-dimethylethyl)-; (25979-07-1)
Methylmagnesium bromide: Magnesium, bromomethyl-; (75-16-1)
Borane-tetrahydrofuran: Boron, trihydro(tetrahydrofuran)-, (14044-65-6)
(*tert*-Butyldimethylphosphine)trihydroboron: Boron, [(1,1-dimethylethyl)dimethyl-phosphine]-trihydro-; (203000-43-5)
sec-Butyllithium: Lithium, (1-methylpropyl)-; (598-30-1)
Copper (II) chloride: Copper chloride; (7447-39-4)
(-)-Sparteine: 7,14-Methano-2*H*,6*H*-dipyrido[1,2-a:1',2'-e][1,5]diazocine, dodecahydro-, (7*S*,7a*R*,14*S*,14a*S*)-; (90-39-1)
Trifluoromethylsulfonic acid: Methanesulfonic acid, trifluoro-; (1493-13-6)
Potassium hydroxide; (1310-58-3)
(Norbornadiene)rhodium chloride dimer; Rhodium, bis[(2,3,5,6-η)-bicyclo[2.2.1]hepta-2,5-diene]di-μ-chlorodi-; (12257-42-0)
Rhodium(1+), [(2,3,5,6-η)-bicyclo[2.2.1]hepta-2,5-diene]bis(methyldiphenylphosphine)-, tetrafluoroborate(1-); (34664-31-8)

PREPARATION OF HEXAKIS(4-BROMOPHENYL)BENZENE (HBB)

[1,1':2',1''-Terphenyl, 4,4''-dibromo-3',4',5',6'-tetrakis(4-bromophenyl)-]

Neat Br_2, 22°C

HPB HBB

Submitted by Rajendra Rathore[1] and Carrie L. Burns.
Checked by Scott E. Denmark and Shinji Fujimori.

1. Procedure

Hexakis(4-bromophenyl)benzene. A 500-mL, three-necked flask, equipped with a mechanical stirrer (fitted with an 11 cm Teflon paddle), a septum and an outlet adapter connected *via* rubber tubing to a pipette which is immersed in 250 mL of a 10% aqueous sodium hydroxide solution in a 500-mL Erlenmeyer flask is charged with 70 mL of bromine (1.37 mol) (Note 1). The septum is replaced with a 250-mL powder addition funnel charged with 26.7 g (50 mmol) of hexaphenylbenzene (HPB) (Note 2). The flask is placed in a water bath at ambient temperature to control the heat evolved from the reaction. To the slowly stirred bromine, hexaphenylbenzene is added slowly over 1 hour (Note 3). The reaction starts immediately as judged by an evolution of gaseous hydrobromic acid (Note 4). After the addition of hexaphenylbenzene is complete, the dark-orange slurry is stirred for an additional 20 min (Note 5).

The bromine slurry of resulting product is carefully poured into 500 mL of pre-chilled (approx. –78°C) ethanol in a 1-L Erlenmeyer flask with stirring by a magnetic stir bar (Note 6). To the three-necked flask is added cold (–78°C) ethanol (2 x 100 mL) and the remaining precipitate is transferred to the Erlenmeyer flask. The suspension of the product in ethanol is allowed to warm to room temperature over 2 hrs. with stirring, and the suspension is filtered using a Büchner funnel. The pale yellow precipitate is washed with ethanol (50 mL), aqueous sodium bisulfite (5%, 100 mL), and ethanol (2 x 50 mL) successively. After being dried overnight *in vacuo* (0.5 mm) at room temperature to a constant weight, 47.8 g (96% yield) of hexakis(4-bromophenyl)benzene (HBB) is obtained with greater than >95% purity as judged by ^{1}H NMR spectroscopy.

The precipitated product is sufficiently pure for most purposes; however, it can be further purified by re-precipitation from tetrahydrofuran. Thus, 10 g of HBB was dissolved in 350 mL of refluxing tetrahydrofuran (Note 7) in a beaker. Upon slow evaporation at room temperature, the solution yields a colorless (microcrystalline) precipitate (9.4 g) of hexakis(4-bromophenyl)benzene (HBB) (Notes 8 and 9).

2. Notes

1. Bromine was obtained from Aldrich Chemical Co. was used as received.

2. Hexaphenylbenzene (HPB) was obtained following the *Organic Syntheses* procedure (Fieser, L, F. *Org. Synth.*, **1973**, *46*, 44;*CV 5*, 604). Commercially available HPB (Aldrich Chemical Co.) can also be used.

3. The reaction is carried out without added solvent and thus for a thorough mixing of reagents an excess of bromine is required for a complete conversion of HPB to HBB. It is critical to maintain the reaction mixture as a slurry for complete conversion.

4. Gaseous HBr was trapped in an aqueous solution of sodium hydroxide.

5. At this point the HBr evolution completely ceases.

6. The slurry should be poured into cold ethanol (-78°C) to prevent an exothermic reaction between excess bromine and ethanol.

7. Tetrahydrofuran (Optima grade) was obtained from Fischer Inc. and was used as received.

8. The spectral data for analytically pure HBB: mp 358-359°C; ^{1}H NMR ($CDCl_3$) δ: 6.61 (d, *J*=8.5 Hz, 12 H), 7.06 (d, *J*=8.6 Hz, 12 H); ^{13}C NMR ($CDCl_3$) δ: 120.3, 130.5, 132.6, 138.4, 139.6. Anal. Calcd for $C_{42}H_{24}Br_6$: C, 50.04; H, 2.40; Br, 47.56. Found: C, 49.92; H, 2.27.

9. Crystallization can be performed in refrigerator (-15°C) over 3 days to provide larger size crystals.

Waste Disposal Information

All hazardous materials should be handled and disposed of in accordance with "Prudent Practices in the Laboratory"; National Academy Press; Washington, DC, 1995.

3. Discussion

The hexaphenylbenzene core is being extensively investigated as a platform for the preparation of nanometer-size macromolecules and supramolecular assemblies owing to their importance as materials that can be used as molecular devices such as sensors, switches, ferromagnets, and other electronic and optoelectronic devices.[2-5] As we recently demonstrated, the elaboration of the hexaphenylbenzene core can be readily achieved using HBB for the preparation of a hexacation-radical salt[6] for use as a (multi)electron-transfer catalyst in a variety of organic and organometallic transformations.[7] The hexaphenylbenzene core is also being utilized for the preparation of well-defined graphite-like structures by Müllen and others and the progress in the area has been reviewed in two recent *Chemical Review* articles.[8,9]

There is one reported procedure for the preparation of HBB, which utilizes the Diels-Alder approach using tetrakis(4-bromophenyl)cyclopentadienone and bis(4-bromophenyl)acetylene (also known as 4,4'-dibromotolan) as the starting materials.[10] Both of these starting materials are prepared *via* multi-step syntheses. We have also discovered that trimerization of 4,4'-dibromotolan using bis(acetonitrile)palladium dichloride affords HBB in fair yield.[11]

A similar bromination procedure (as described above for the preparation of HBB) can be employed for the preparation of tetrakis(4-bromophenyl)methane using tetraphenylmethane and bromine.[6]

The method described here for the preparation of HBB is essentially a detailed description of our recently published procedure[6] using hexaphenylbenzene and neat bromine. It is believed that the ready availability of HBB from hexaphenylbenzene will facilitate the synthesis of a variety of materials, which were otherwise not readily accessible.[12]

1. Department of Chemistry, Marquette University, P.O. Box 1881, Milwaukee, WI 53201-1881.
2. Lambert, C.; Noll, G. *Angew. Chem. Int. Ed. Engl.* **1998,** *37*, 2107.

3. Praefcke, K.; Khone, B.; Singer, D. *Angew. Chem. Intl. Ed. Engl.* **1990,** *29*, 177.
4. Laschewsky, A. *Angew. Chem. Intl. Ed. Engl.* **1989**, *28*, 1574.
5. Kobayashi, K.; Shiraska, T.; Horn, E.; Furukawa, N. *Tetrahedron Lett.* **2000**, *41*, 89 and references therein.
6. Rathore, R.; Burns, C.L.; Deselnicu, M. I. *Org. Lett.* **2001**, *3*, 2887.
7. Connelly, N.G.; Geiger, W.E. *Chem. Rev.* **1996,** *96*, 877 and references therein.
8. Watson, M.D.; Fechtenkotter, A.; Mullen, K. *Chem. Rev.* **2001**, *101*, 1267 (review) and references therein.
9. Berresheim, A.J.; Muller, M.; Mullen, K. *Chem. Rev.* **1999**, *99*, 1747 (review) and references therein.
10. Broser, W.; Siegle, P.; Curreck, H. *Chem. Ber.* **1968**, *101*, 69-83.
11. Rathore, R. (unpublished results).
12. Rathore, R., Burns, C. L., Guzei, I. A. *J. Org. Chem.* **2004**, *69*. 1524-1530.

Appendix
Chemical Abstracts Nomenclature (Registry Number)

Hexaphenylbenzene: 1,1':2',1"-Terphenyl, 3',4',5',6'-tetraphenyl-; (992-04-1)

Hexakis(4-Bromophenyl)benzene: 1,1':2',1"-Terphenyl, 4,4"-dibromo-3',4',5',6'-tetrakis(4-bromophenyl)-; (19057-50-2)

Sodium bisulfite: Sulfurous acid, monosodium salt; (7631-90-5)

Bromine; (7726-95-6)

AN EFFICIENT, HIGHLY DIASTEREO- AND ENANTIOSELECTIVE HETERO-DIELS-ALDER CATALYST. PREPARATION OF (2*S*,6*R*)-6-(*tert*-BUTYLDIMETHYL-SILYLOXYMETHYL)-2-METHOXY-2,5-DIHYDROPYRAN

(Silane, [[(2*R*,6*S*)-3,6-dihydro-6-methoxy-2H-pyran-2-yl]methoxy]-(1,1-dimethylethyl)dimethyl)-

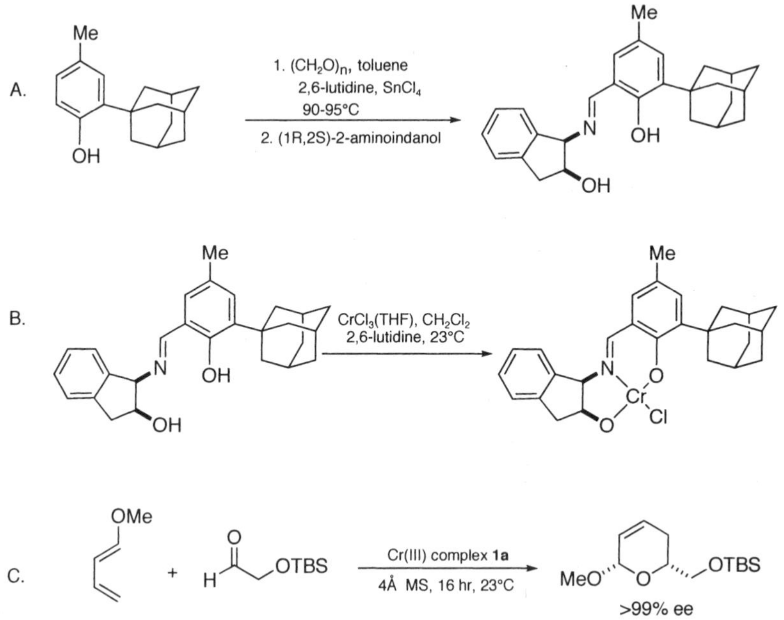
A.
Me
OH
1. (CH2O)n, toluene
2,6-lutidine, SnCl4
90-95°C
2. (1R,2S)-2-aminoindanol
Me
N
OH
OH
B.
Me
N
OH
OH
CrCl3(THF), CH2Cl2
2,6-lutidine, 23°C
Me
N
O
Cr
Cl
O
C.
OMe
O
H
OTBS
Cr(III) complex 1a
4Å MS, 16 hr, 23°C
MeO
O
OTBS
>99% ee

Submitted by David E. Chavez and Eric N. Jacobsen.[1]
Checked by E. J. J. Grabowski and Michele Kubryk.

1. Procedure

A. *(1R,2S)-1-[3-Adamantyl)-2-hydroxy-5-methylbenzylidenamino]indan-2-ol.* An oven-dried, 300-mL, three-necked, round-bottomed flask is equipped with a magnetic stir bar, fitted with a reflux condenser and thermometer, and purged with a nitrogen atmosphere by means of an inlet fitted to the condenser. The flask is charged with 2-adamantyl-4-methylphenol (12.1 g, 50.0 mmol, 1 eq) (Note 1), 110 mL of freshly distilled toluene (Note 2), and 2,6-lutidine (4.28 g, 4.67 mL, 40.00 mmol, 0.8 eq); the open neck of the flask is capped with a septum. Neat stannic chloride

($SnCl_4$) (2.60 g, 1.17 mL, 10.00 mmol, 0.2 eq) is added by syringe over 10 min (Note 3). The solution turns pale yellow in color, and a pale yellow precipitate is also observable. The mixture is allowed to stir at room temperature for 20 min, then the septum is removed and solid paraformaldehyde (6.00 g, 200 mmol) is added in one portion against a gentle nitrogen counterflow (Note 4). The mixture is stirred an additional 10 min, the nitrogen inlet is replaced with a nitrogen balloon, the reaction flask is placed in a 90-95°C bath, and heating is maintained at this temperature for 6 hr. The reaction mixture is then allowed to cool to room temperature and filtered through a pad of premixed Celite® and silica gel (1:1, 12 g). The filter pad is washed with ethyl acetate (200 mL), and the combined organic filtrates are washed with water (350 mL), 1N HCl (350 mL), and brine (350 mL), and then dried over anhydrous Na_2SO_4. Concentration is effected by rotary evaporation, followed by removal of trace solvent on a high vacuum pump (0.5 mm) (13.4 g crude, 99.5%) (Note 5). Absolute ethanol (200 mL) is added and the mixture is heated gently until complete dissolution occurs (Note 6). (1*R*,2*S*)-1-Amino-2-indanol (7.83 g, 52.50 mmol, 1.05 equiv.) (Note 2) is added in one portion. The reaction mixture is then heated at 80°C for 45 min, cooled to room temperature, and allowed to stand for 3-5 hours. The yellow solid product is isolated filtration, washed with cold ethanol (50 mL), and dried in the air (15.1 g, 75.2% over 2 steps) (Note 7).

B. *Chromium(III) Cl complex (**1a**)*. To a 200-mL round-bottomed flask is added chromium(III) chloride-tetrahydrofuran complex (1:3) (2.80 g, 7.48 mmol, 1 equiv.) and (1*R*,2*S*)-1-[3-adamantyl)-2-hydroxy-5-methylbenzylidenamino]indan-2-ol (3.00 g, 7.48 mmol, 1 equiv.). The reaction mixture is placed under a nitrogen atmosphere, and 60 mL of dichloromethane (CH_2Cl_2) is added followed by dropwise addition of 2,6-lutidine (1.74 mL, 14.96 mmol, 2 equiv). The solution is stirred for 3 hr, diluted with CH_2Cl_2 (300 mL), and washed with water (3 x 180 mL), then brine (180 mL) (Note 8). The organic phase is dried over anhydrous Na_2SO_4, filtered, and concentrated by rotary evaporation. The resulting solid is triturated with acetone (10 mL), filtered, washed with an additional portion of acetone (10 mL), and air-dried to give the chromium complex (**1a**) as a brown solid (2.3 g). Water (2 mL) is added to the filtrate (Note 9) and the solution allowed to stand uncovered at 23°C overnight. The resulting precipitate is filtered and washed with cold acetone to give an additional 600-800 mg of the chromium complex (**1a**) (combined yield 2.9-3.1 g, 80 – 85%) (Notes 10, 11).

C. *(2S,6R)-6-(tert-Butyldimethylsilyloxymethyl)-2-methoxy-2,5-dihydro-pyran.* 1-Methoxybutadiene (2.40 g, 2.89 mL, 28.7 mmol, 1.11 equiv.) is added dropwise to a stirring mixture of (*tert*-butyldimethylsilyloxy)acetaldehyde (90%, 5.00 g, 5.46 mL, 25.8 mmol, 1 equiv.), (1*R*, 2*S*) chromium(III) chloride complex (**1a**) (200 mg, 0.19 mmol, 1.5 mol% (Note 12) and 4Å molecular sieves (Note 13) under N_2 at 0°C. The reaction mixture is allowed to stir at 0°C for 1 hr and then warmed to room temperature and allowed to stir for an additional 16 hr. Distillation of this mixture (Kügelrohr, 110°C, 0.5 mm) affords the cycloadduct (6.0 g, 90%) as a colorless oil (Note 14) in >99% ee (Note 15).

2. Notes

1. The purity of the 2-adamantyl-4-methylphenol is important; in particular, the material should be free of 2,6-diadamantyl-4-methylphenol.

2. All reagents were obtained from commercial suppliers (Acros, Aldrich Chemical Company, Inc., or Strem Chemicals, Inc.). Toluene was distilled from sodium, and dichloromethane was distilled from calcium hydride. All other reagents were used as received without further purification.

3. The use of a syringe containing a teflon plunger prevents clogging during the addition of $SnCl_4$.

4. Caution must be taken to prevent the fluffy solid paraformaldehyde from dispersing outside of the flask during this addition process.

5. This procedure for the synthesis of 2-adamantyl-5-methylsalicylaldehyde is a modification of the method reported by Casiraghi.[2] The aldehyde can be recrystallized from hexanes, but purification is not essential for successful formation of the Schiff base. The purified aldehyde has the following spectral and physical properties: mp 151.5-152°C; IR (KBr) 3200-2500, 1649, 1607, 1524, 1447, 1416, 1356, 1312, 1244, 1221, 1163, 1105, 1084, 1040, 963, 864 cm^{-1}; ^{1}H NMR (500 MHz, $CDCl_3$) δ: 1.78 (s, 6 H), 2.08 (s, 3 H), 2.12 (s, 6 H), 2.31 (s, 3 H), 7.14 (d, J = 1.5 Hz, 1 H), 7.26 (d, J = 1.5 Hz, 1 H), 9.8 (s, 1 H), 11.65 (s, 1 H); ^{13}C NMR (100 MHz, $CDCl_3$) δ: 20.5, 28.9, 36.9, 40.1, 120.3, 128.2, 131.2, 135.4, 138.1, 159.3, 197.1; Anal. Calcd for $C_{18}H_{22}O_2$: C, 79.96; H, 8.20. Found: C, 79.70; H, 8.16.

6. The aldehyde is observed to dissolve completely between 60-70°C.

7. The product exhibits the following physical and spectroscopic properties: mp 219-221°C; $[\alpha]^{26}_{D}$ +70.0 (c .100, THF); IR (KBr disk) 3584, 2905, 2849, 1624, 1597cm^{-1}; ^{1}H NMR (500 MHz, DMSO-d_6) δ: 1.69 (m, 6 H), 1.99 (m, 3 H), 2.05 (m, 6 H), 2.23 (s, 3 H), 2.95 (dd, J = 6.0, 15.5 Hz, 1 H), 3.11 (dd, J = 6.1, 15.5 Hz, 1 H), 4.54 ('q', J = 5.7 Hz, 1 H), 4.73, (d, J =

5.5 Hz, 1 H), 5.23, (d, J = 4.9 Hz, 1 H), 7.01 (s, 1 H), 7.09 (s, 1 H), 7.18-7.31 (m, 4 H), 8.61 (s, 1 H), 10.94 (s, 1 H); ^{13}C-NMR (125 MHz, DMSO-d_6) δ: 20.2, 28.3, 36.2, 36.5, 39.0, 39.7, 73.9 (2 carbons), 118.2, 124.7, 125.0, 125.7, 126.6, 127.4, 127.9, 129.6, 130.0, 136.4, 141.0, 142.0, 158.5, 166.5; HRMS (CI, NH_3) *m/z* calcd for $C_{27}H_{35}NO_2(M)^+$ 401.2355, found 401.2341.

8. The water washes should be carried out with gentle shaking in order to avoid formation of intractable emulsions.

9. If partial concentration occurs during filtration, the filtrate should be diluted with acetone prior to addition of water such that the total volume is 20 mL. Upon addition of water, a small amount of precipitate may form. This should be redissolved by gently warming the solution or by addition of a minimal amount of acetone.

10. X-Ray quality crystals are obtained by recrystallization from acetone/water. The solid state structure of complex **1** is that of a dimer bearing a bridging water molecule and one terminal water molecule on each metal center.[3] This dimeric complex exhibits the following spectral properties: IR (KBr): 3414, 2903, 2847, 1618, 1537, 1433, 1340, 1305, 1228, 1168, 1078 cm^{-1}. LRMS (FAB): calcd for dimer $C_{54}H_{68}Cl_2N_2O_7Cr_2$, $(M\text{-}2Cl\text{-}2H_2O)^+$, 920, found 919. A dehydrated sample suitable for elemental analysis was prepared as follows: Chlorotrimethylsilane (39.0 μL, 0.31 mmol) was added to a solution of Cr(III)Cl complex (50.0 mg, 0.048 mmol) in dry *tert*-butyl methyl ether (2 mL). The mixture was stirred for 2 hr under nitrogen to give a green precipitate. The mixture was concentrated *in vacuo*, suspended in dry *tert*-butyl methyl ether (2 mL), filtered and the residue washed with dry *tert*-butyl methyl ether. The residue was then dried under high vacuum (0.5 mm). Anal. Calcd for $[C_{27}H_{29}ClCrNO_2 + 2HCl]$: C, 57.92; H, 5.58; Cr, 9.29; N 2.50. Found: C, 57.49; H, 5.73; Cr, 9.00; N, 2.48.

11. For certain applications (see, for example, the first entry in Table 1), superior results in HDA reactions are obtained with catalyst **1b**, wherein the chloride counterion of **1a** is replaced with SbF_6. Preparation of catalyst **1b** is achieved as follows: A flame-dried, 50-mL, foil wrapped round-bottomed flask equipped with a stirbar was charged with complex **1a** (100 mg, 0.97 mmol, 1 equiv) and silver hexafluoroantimonate (66.8 mg, 0.19 mmol, 2 equiv). The flask was placed under a nitrogen atmosphere, *tert*-butyl methyl ether (30 mL) was added, and the mixture allowed to stir for 3 hr. The reaction mixture was then filtered through Celite® and the isolated solids are washed with *tert*-butyl methyl ether (20 mL). The filtrates were combined and concentrated by rotary evaporation to afford the desired SbF_6 complex **1b** as a brown solid (165 mg). IR (KBr) 3378, 2973, 2905, 1615,

1538, 1229, 1069 cm^{-1}. LRMS (FAB) *(m/z)* calcd for $C_{27}H_{35}CrNO_2$ $(M)^+$ 451; found 451; calcd for $2[C_{27}H_{35}CrNO_2]$, $(2M)^+$, 902; found 902; calcd for 2 $C_{27}H_{35}CrNO_2$ + H_2O, $(2M + H_2O)^+$, 920; found 921.

12. The catalyst loading was calculated based on the number of equivalents of chromium relative to the limiting aldehyde substrate.

13. The molecular sieves (1.6 mm pellets) are powdered with a mortar and pestle and activated in a vacuum oven (130°C) overnight before use. Alternatively, commercially available finely powdered 4Å molecular sieves (<5 micron) may be used.

14. The product has the following spectral and physical properties: $[\alpha]^{26}_D$ +55.3 (c 1.14, $CDCl_3$); R_f = 0.70 (1:1 ether/hexanes); IR (thin film) 2955, 2934, 2888, 2858, 1471, 1400, 1339, 1255, 1204, 1129, 1112, 1080, 1057 cm^{-1}; ^1H NMR (500 MHz, $CDCl_3$) δ: 0.07 (s, 6 H), 0.89 (s, 9 H), 2.08 (m, 2 H), 3.47 (s, 3 H), 3.65 (dd, J = 6.5, 10.4 Hz, 1 H), 3.76 (dd, J = 5.6, 10.4 Hz, 1 H), 3.85 ('q', J = 6.3 Hz, 1 H), 5.02 (m, 1 H), 5.65 ('dq', J = 3.7, 10.2 Hz, 1 H), 5.97 ('dq', J = 5.3, 10.2 Hz, 1 H); ^{13}C NMR (125 MHz, $CDCl_3$) δ: 5.2, 5.3, 18.4, 25.9, 26.8, 55.2, 65.5, 72.6, 97.7, 127.0 128.5; HRMS *(m/z)* (Cl) calcd for $C_{13}H_{30}NO_3Si$ $(M+NH_4)^+$ 276.1995, found 276.2003.

15. Enantiomeric excess was determined by GC analysis following conversion to (*R*)-6-(*tert*-butyldimethylsilyoxymethyl)-5,6-dihydropyran-2-one, according to the following procedure: Pyridinium dichromate (1.04 g, 2.75 mmol) was added to a solution of the acetal (256 mg, 1.38 mmol) and acetic acid (3 mL) in CH_2Cl_2 (20 mL) at 23°C. The mixture was stirred overnight, diluted with 1:1 ether/hexanes (20 mL), and filtered through a pad of $MgSO_4$. The residue remaining in the reaction flask was washed thoroughly with 1:1 ether/hexanes (4 x 20 mL) and the extracts were filtered. The combined filtrates were filtered once more through a fresh pad of $MgSO_4$ and concentrated *in vacuo*. Kügelrohr distillation (210-220°C, 10 mm) afforded the product lactone (267 mg, 57.0%). GC analysis using a commercial chiral column (Cyclodex β. 135°C, isothermal) revealed the product to be in >99% ee (t_R(major) = 50.23 min). $[\alpha]^{26}_D$ +79 (c 1.00, $CDCl_3$). R_f = 0.17 (10% ether/hexanes). IR (thin film) 2955, 2930, 2859, 1732, 1471, 1407, 251, 1136, 1093, 1043 cm^{-1}. ^1H NMR (500 MHz, $CDCl_3$) δ: 0.06 (s, 6 H), 0.87 (s, 9 H), 2.40 ('dt', J = 4.6, 18.6 Hz, 1 H), 2.51 (ddd, J = 2.6, 11.1, 18.6 Hz, 1 H), 3.78 (dd, J = 5.4, 10.9 Hz, 1 H), 3.80 (dd, J = 4.64, 10.9 Hz, 1 H), 4.45 (dddd, J = 4.4, 4.6, 5.4, 11.1 Hz, 1 H), 5.99 (d, J = 9.7 Hz, 1 H), 6.89 (ddd, J = 2.6, 5.8, 9.7 Hz, 1 H). ^{13}C NMR (125 MHz, $CDCl_3$) δ: -5.4, 18.3, 25.8, 64.2, 77.8, 121.1, 145.0, 163.9. HRMS (Cl) *m/z* calcd for $C_{12}H_{26}NO_3Si$ $(M+NH_4)^+$ 260.1682, found 260.1679.

Waste Disposal Information

All hazardous materials should be handled and disposed of in accordance with "Prudent Practices in the Laboratory"; National Academy Press; Washington, DC, 1995.

3. Discussion

This procedure describes a practical synthesis of the chiral tridentate Schiff base complex **1a**, and the use of this complex to catalyze an efficient, highly diastereo- and enantioselective hetero-Diels-Alder (HDA) reaction. The unique characteristic of this catalyst, and the derived SbF_6 complex **1b** (see Note 11), lies in their demonstrated ability to promote asymmetric hetero-Diels-Alder reactions between aldehydes and dienes bearing a single oxygen substituent.[4] Reactions proceed generally with excellent diastereo- and enantioselectivity, and provide access to enantiomerically enriched dihydropyran derivatives from simple achiral starting materials (Table 1). This HDA methodology has already been showcased in several natural product syntheses.[5] More recently, the same catalyst system has been applied to highly enantioselective inverse demand hetero-Diels-Alder reactions between conjugated aldehydes and ethyl vinyl ether.[3]

The method for the synthesis of complex **1a** described herein represents a significant improvement over the procedure first reported in 1999.[4] The use of air- and moisture-sensitive $CrCl_2$ is now avoided, and the necessity of conducting the metal-insertion step in a glove box is thereby precluded. Instead, the use of the ($Cr(III)Cl_3 \cdot [C_4H_8O]_3$) complex allows the reaction to be conducted in a fume hood. Additionally, the procedure for the formylation of 2-adamantyl-4-methylphenol has been adapted such that purification of the resulting aldehyde by recrystallization is no longer necessary. Finally, and perhaps most important, catalysts prepared by the new procedure displays measurably higher enantioselectivity in a variety of HDA reactions.[3]

Table 1

Aldehyde	Diene	Product	Cat	ee (%)	Yield (%)	Ref
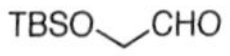 TBSO CHO	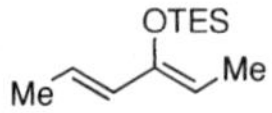 OTES Me Me	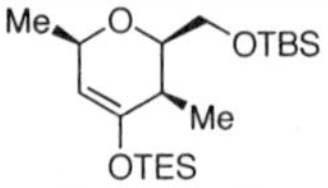 Me O OTBS Me OTES	**1a**	99	90	4
			1b	>99	97	4
PhCHO	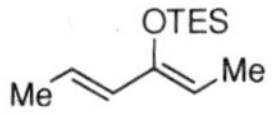 OTES Me Me	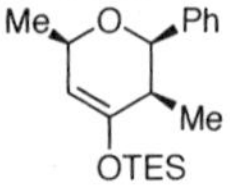 Me O Ph Me OTES	**1b**	90	72	4
n-C_5H_{11}CHO	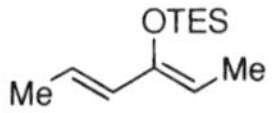 OTES Me Me	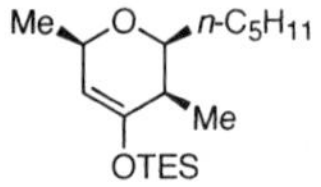 Me O n-C_5H_{11} Me OTES	**1b**	98	85	4
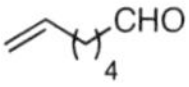 CHO (4)	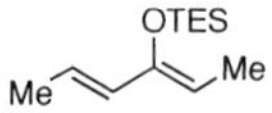 OTES Me Me	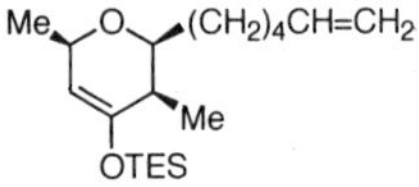 Me O $(CH_2)_4CH{=}CH_2$ Me OTES	**1b**	98	85	4
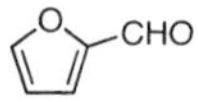 O CHO	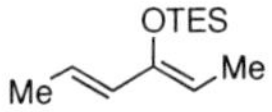 OTES Me Me	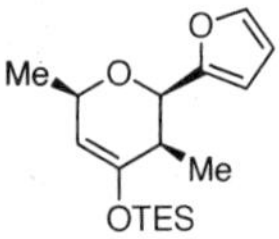 Me O O Me OTES	**1b**	95	77	4
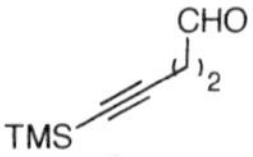 CHO (2) TMS	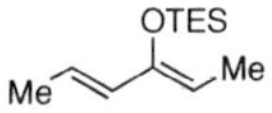 OTES Me Me	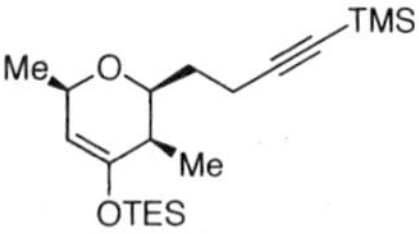 Me O TMS Me OTES	**1b**	95	92	5a
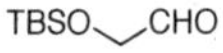 TBSO CHO	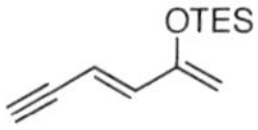 OTES	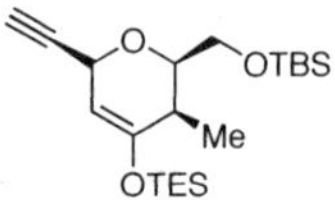 O OTBS Me OTES	**1b**	98	61	5a
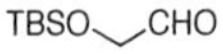 TBSO CHO	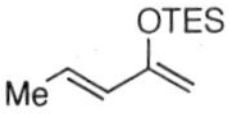 OTES Me	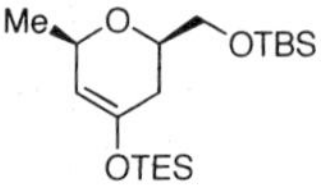 Me O OTBS OTES	**1b**	98	78	4
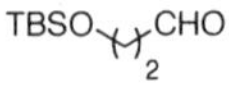 TBSO (2) CHO	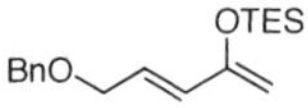 OTES BnO	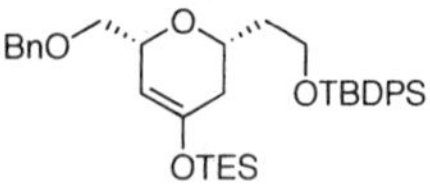 BnO O OTBDPS OTES	**1a**	97	64	5c
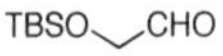 TBSO CHO	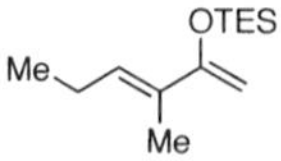 OTES Me Me	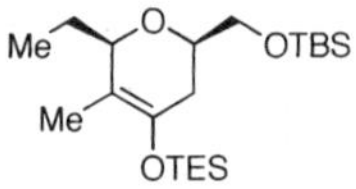 Me O OTBS Me OTES	**1a**	>99	87	5c
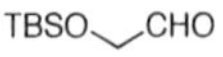 TBSO CHO	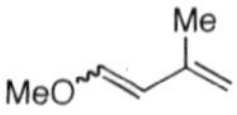 Me MeO	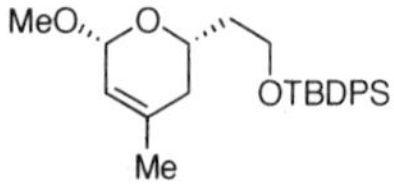 MeO O OTBDPS Me	**1a**	>95	71	5e
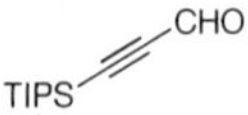 CHO TIPS	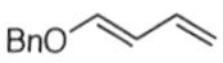 BnO	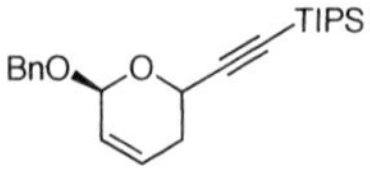 BnO O TIPS	**1a**	94	80	5b, 3

The hetero-Diels-Alder reaction illustrated in this procedure utilizes commercially available 1-methoxy-1,3-butadiene and (*t*-butyldimethylsilyloxy)acetaldehyde. The reaction is carried out with 1.5 mol% catalyst under solvent-free conditions. The dihydropyran is isolated in 90% yield, >97:3 dr, and >99% ee by direct distillation of the reaction mixture. The product can be oxidized to the corresponding lactone readily and in one step providing efficient access to a substructure that occurs in several interesting natural products (i.e., fostriecin[5b], callystatin A[6a], ratjadone[6b]).

1. Department of Chemistry and Chemical Biology, Harvard University, Cambridge, MA 02138.
2. Casiraghi, G.; Casnati, G.; Puglia, G.; Sartori, G.; Terenghi, G. *J. Chem. Soc., Perkin Trans.* **1980**, 1862.
3. Gademann, K.; Chavez, D. E.; Jacobsen, E. N. *Angew. Chem. Int. Ed.* **2002**, *41*, in press.
4. Dossetter, A. G.; Jamison, T. F.; Jacobsen, E. N. *Angew. Chem. Int. Ed.* **1999**, *38*, 2398.
5. (a) FR901464: Thompson, C. F.; Jamison, T. F.; Jacobsen, E. N. *J. Am. Chem. Soc.* **2000**, *122*, 10482; *J. Am. Chem. Soc.* **2001**, *121*, 9974 (b) Fostriecin: Chavez, D. E.; Jacobsen, E. N. *Angew. Chem., Int. Ed. Engl.* **2001**, *40*, 3667; c) Ambruticin: Liu, P.; Jacobsen, E. N. *J. Am. Chem. Soc.* **2001**, *123*, 10772; (d) Apicularin A: Bhattacharjee, A.; De Brabander, J. K. *Tetrahedron Lett.* **2000**, 41, 8069. (e) Laulilamide: Paterson, I.; De Savi, C.; Tutdge, M. *Org. Lett.* **2001**, *3*, 3149.
6. (a) Callystatin A: Crimmins, M. T.; King, B. W. *J. Am. Chem. Soc.* **1998**, *120*, 9084; (b) Ratjadone: Christman, M.; Bhatt, U.; Quitschalle, E.: Claus, E.; Kalesse, M. *Angew. Chem. Int. Ed.* **2000**, *39*, 4364.

Appendix
Chemical Abstracts Nomenclature (Registry Number)

2-(1-Adamantyl)-4-methylphenol: Phenol, 4-methyl-2-tricyclo[3.3.1.13,7]-dec-1-yl-; (41031-50-9)

(1*R*,2*S*)-1-Aminoindan-2-ol: 1*H*-Inden-2-ol, 1-amino-2,3-dihydro-,(1*S*-*cis*); (126456-43-7)

(1*R*,2*S*)-1-[(3-Adamantyl)-2-hydroxy-5-methylbenzylidenamino]indan-2-ol: 1*H*-Inden-2-ol, 2,3-dihydro-1-[[(2-hydroxy-5-methyl-3-tricyclo[3.3.1.13,7]dec-1-decylphenyl)methylene]amino]-, (1*R*,2*S*)-; (231963-92-1)

Chromium(III) Cl Complex: Chromium, chloro[(1*R*,2*S*)-2,3-dihydro-1-[[[2-(hydroxy-κ*O*)-5-methyl-3-tricyclo[3.3.1.13,7]dec-1-ylphenyl]methylene]amino-κ*N*]-1*H*-indene-2-olato-(2-)-κ*O*],(SP-4-4); (231963-76-1)

1-Methoxy-1,3-butadiene: 1,3-Butadiene, 1-methoxy-; (3036-66-6)

(*tert*-Butyldimethylsilyloxy)acetaldehyde: Acetaldehyde, [[(1,1-dimethylethyldimethylsilyl]oxy]-; (102191-92-4)

(2*S*,6*R*)-6-(*tert*-Butyldimethylsilyloxymethyl)-2-methoxy-2,5-dihydropyran: Silane, [[(2*R*,6*S*)-3,6-dihydro-6-methoxy-2*H*-pyran-2-yl]methoxy](1,1-dimethylethyl)dimethyl-; (231963-89-6)

LIPASE-CATALYZED RESOLUTION OF 4-TRIMETHYLSILYL-3-BUTYN-2-OL AND CONVERSION OF THE (*R*)-ENANTIOMER TO (*R*)-3-BUTYN-2-YL MESYLATE AND (*P*)-1-TRIBUTYLSTANNYL-1,2-BUTADIENE

[3-Butyn-2-ol, 4-(trimethylsilyl)-, (2*R*)- and Stannane, 1,2-butadienyltributyl-, *(P)*-]

A. Me_3Si (OH, Me) —Amano Lipase AK, pentanes→ Me_3Si (*S*) OH, H, Me **1** + Me_3Si (*R*) OAc, Me, H **2**

1. (succinic anhydride), Et_3N, DMAP 2. $NaHCO_3$ → Me_3Si (*S*) O, O, CO_2Na, H, Me —HCl, H_2O→ Me_3Si (*S*) O, O, CO_2H, H, Me **3**

B. Me_3Si (*R*) OAc, Me, H **2** —DIBAL-H, hexanes→ Me_3Si (*R*) OH, Me, H **4**

C. Me_3Si (*S*) O, O, CO_2H, H, Me **3** —DIBAL-H, hexanes→ Me_3Si (*S*) OH, H, Me **1**

D. Me_3Si (*R*) OH, Me, H **4** —MsCl, Et_3N→ Me_3Si (*R*) OMs, Me, H **5**

E. Me_3Si (*R*) OMs, Me, H **5** —K_2CO_3, MeOH→ H (*R*) OMs, Me, H **6**

F. H (*R*) OMs, Me, H **6** —a) LDA, Bu_3SnH; then $Me_2S{\cdot}CuBr$ b) Addition of **6**→ H, Bu_3Sn (*P*) Me, H **7**

Submitted by James A. Marshall and Harry Chobanian.[1]
Checked by Peter Wipf and Joshua Pierce.

1. Procedure

A. *(R)-4-Trimethylsilyl-3-butyn-2-yl acetate (2) and (S)-4-Trimethylsilyl-3-butyn-2-yl succinate* (**3**). An oven dried, 1-L, round-bottomed flask, equipped with a large magnetic stir bar and rubber septum with nitrogen inlet, is charged with racemic 4-trimethylsilyl-3-butyn-2-ol (**1**) (10.0 g, 69.9 mmol) (Note 1) and 250 mL of pentane (Note 2). To this stirred solution is added Amano Lipase AK (Note 3) (2.00 g), freshly distilled vinyl acetate (50 mL), and pulverized, activated 4Å molecular sieves (1 g). The mixture is stirred at room temperature for 72 hr at which point analysis by GC (Note 4) indicated the reaction had proceeded to 50% completion. The mixture is filtered through a medium porosity sintered glass funnel, washed with additional pentane, and concentrated via rotary evaporation affording 11.5 g of a nearly 1:1 mixture of alcohol and acetate by ^{1}H NMR analysis. To this mixture in 50 mL of THF is added Et_3N (9.1 mL, 65.0 mmol), DMAP (92 mg, 0.75 mmol) and succinic anhydride (4.15 g, 41.1 mmol) successively. The mixture is heated to reflux for 4 hr, cooled, and quenched with 40 mL of sat. aq. $NaHCO_3$ to adjust the pH to ≥9. The solution is stirred vigorously for 1 hr and diluted with ethyl acetate (EtOAc) (75 mL). The EtOAc solution is separated, washed with 10% HCl (200 mL) and brine (200 mL), dried over $MgSO_4$, filtered, and concentrated by rotary evaporation. The resulting oil is purified by bulb-to-bulb distillation (65°C at 0.5 mmHg) (Note 5) to yield 6.10-6.11 g (94%) of (*R*)-4-trimethylsilyl-3-butyn-2-yl acetate (**2**) as a clear oil (Note 6). The aqueous phase is carefully acidified with 12 M HCl (~5 mL) to pH ~ 1 and extracted with Et_2O (4 x 150 mL). The combined Et_2O extracts are dried over $MgSO_4$, filtered, and concentrated by rotary evaporation affording 8.47–8.48 g (99%) of (*S*)-4-trimethylsilyl-3-butyn-2-yl succinate (**3**) as a light yellow oil which solidified upon cooling to 0°C. The acid is carried on without further purification (Note 7).

B. *(R)-4-Trimethylsilyl-3-butyn-2-ol* (**4**). To an oven-dried, 100-mL, round-bottomed flask, equipped with a magnetic stir bar and rubber septum with nitrogen inlet is added acetate **2** (6.11 g, 33.2 mmol) in 40 mL of hexanes at -78°C followed by 50 mL of DIBAL-H (1.0 M in hexanes). The solution is stirred for 10 min and poured into a rapidly stirred mixture of 300 mL of aqueous Rochelle's salt and 200 mL of Et_2O. Once the Et_2O layer clarifies, it is separated, dried over $MgSO_4$, filtered, and concentrated by distillation at atmospheric pressure to yield an oil. The oil is purified by bulb-to-bulb distillation (80°C at 0.5 mmHg) to yield 4.27-4.25 g (87%) of (*R*)-4-trimethylsilyl-3-butyn-2-ol (**4**) (Note 8).

C. *(S)-4-Trimethylsilyl-3-butyn-2-ol* (**1**). To an oven-dried, 250-mL, round-bottomed flask, equipped with a magnetic stir bar and rubber septum with nitrogen inlet is added a solution of the foregoing succinate **3** (8.48 g, 34.8 mmol) in 100 mL of CH_2Cl_2 at -78 °C followed by 77 mL of DIBAL-H (1.0 M in hexanes). The solution is stirred for 10 min and the product is isolated as described for (*R*)-4-trimethylsilyl-3-butyn-2-ol (**4**). The resulting oil is purified by bulb-to-bulb distillation (75-80°C at 0.5 mmHg) to yield 4.26-4.28 g (86%) of (*S*)-4-trimethylsilyl-3-butyn-2-ol (**1**) (Notes 9 and 10).

D. *(R)-4-Trimethylsilyl-3-butyn-2-yl mesylate* (**5**). An oven-dried, 500-mL, round-bottomed flask equipped with a magnetic stir bar and rubber septum with nitrogen inlet is charged with (*R*)-4-trimethylsilyl-3-butyn-2-ol (**4**) (4.28 g, 30.1 mmol) and 350 mL of CH_2Cl_2 at -78°C. To this solution is added Et_3N (8.50 mL, 61.0 mmol) followed by methanesulfonyl chloride (5.23 g, 46.0 mmol). The solution is stirred for 1 hr at -78°C before being quenched with 10 mL of sat. aq. $NaHCO_3$ solution. The solution is allowed to warm to room temperature before being concentrated at reduced pressure by rotary evaporation. The residual oil is diluted with distilled H_2O (200 mL) and Et_2O (100 mL). The Et_2O layer is separated, dried over $MgSO_4$, filtered, and concentrated to afford 6.07-6.09 g (91%) of (*R*)-4-trimethylsilyl-3-butyn-2-yl mesylate (**5**), which can be used without further purification (Note 11).

E. *(R)-3-Butyn-2-yl mesylate* (**6**). To a 100-mL round-bottomed flask, equipped with a magnetic stir bar, is added a solution of (*R*)-4-trimethylsilyl-3-butyn-2-yl mesylate (6.09 g, 27.6 mmol) in 20 mL of methanol followed by addition of anhydrous K_2CO_3 (5.76 g, 42.0 mmol). After 15 min, TLC analysis (1:1; EtOAc:hexanes) of the mixture indicates complete cleavage of the TMS group. The solution is diluted with 40 mL of sat NaCl solution and extracted with diethyl ether (Et_2O) (3 x 50 mL). The combined ether layers are washed with distilled H_2O (3 x 50 mL) to remove residual methanol. The ether layer is separated, dried over $MgSO_4$, filtered, and concentrated under reduced pressure (Note 12) to yield 3.48 g (85%) of (*R*)-3-butyn-2-yl mesylate (**6**) as a clear oil that can be used without further purification (Note 13).

F. *(P)-(+)-3-(Tributylstannyl)-1,2-butadiene* (**7**). An oven-dried, 250-mL, round-bottomed flask equipped with a magnetic stir bar and a rubber septum with a nitrogen inlet is charged with diisopropylamine (4.0 mL, 28.0 mmol) and 100 mL of THF. To this solution is added 11.2 mL (28.0 mmol) of *n*-BuLi (2.5 M in hexanes) at 0°C. After 30 min, tributyltin hydride (8.15 g, 28.0 mmol) (Note 14) is added dropwise and the mixture is stirred an additional 30 min. The yellow solution is then cooled to -78°C and 5.76 g (28.0 mmol) of $CuBr{\cdot}SMe_2$ is added portion-wise (Note 15). Once addition is complete, the dark solution is stirred an additional 30 min before (*R*)-3-butyn-2-yl mesylate (**6**) (Note 16) is added. After 15 min, the solution is poured into a rapidly stirred solution of 400 mL of 9:1 sat. aq. NH_4Cl/NH_4OH solution and 300 mL of ether. Once the ether layer clarifies, it is separated, dried over $MgSO_4$, filtered, and concentrated under reduced pressure. The resulting oil is purified by bulb-to-bulb distillation (120°C/ 0.5 mmHg) to yield 5.77-5.79 g (88%) of (*P*)-(+)-3-(tributylstannyl)-1,2-butadiene as a yellow oil (Note 17).

2. Notes

1. 4-Trimethylsilyl-3-butyn-2-ol was obtained from Gelest, Inc. (Submitters) or from Lancaster Research Chemicals (Checkers). All other chemicals were purchased from Aldrich Chemical Company, Inc. and used as received.

2. The selectivity of the resolution diminished at a higher concentration of butynol. A decrease in the volume of pentane from 250 mL to 125 mL led to acetate with er = 88:12 at 50% conversion.

3. Amano Lipase AK from *Pseudomonas fluorescens* was purchased from Aldrich Chemical Company, Inc.

4. The reaction progress was monitored by GC (Carbowax, 110°C, 1°C ramp/min; acetate 5.75 min; alcohol 8.22 min. or β-DEX, 89 °C, 0.2°C ramp/min, alcohol 31.30 min, acetate 32.37 min).

5. Bulb-to-bulb (short-path) distillation was performed with an Aldrich Kugelrohr distillation apparatus.

6. Physical characteristics of (*R*)-2-acetoxy-4-trimethylsilyl-3-butyne (**2**): $[\alpha]_D^{20}$ +119 (c = 2.2, $CHCl_3$); IR (film) cm^{-1}: 2181, 1747; ^1H NMR (300 MHz, $CDCl_3$) δ: 0.15 (s, 9 H), 1.45 (d, *J* = 6.7, 3 H), 2.05 (s, 3 H), 5.44 (q, *J* = 6.7, 1 H). ^{13}C NMR (75 MHz, $CDCl_3$) δ: -0.27, 21.1, 21.4, 60.6, 89.4, 103.5, 169.7. Analysis by GC on a β-Dex column showed a single peak at 32.0 min (80°C, 0.2°C ramp/min). The racemic acetate gave rise to peaks at 31.66 and 32.42 min under these conditions.

7. Spectral analysis for (*S*)-4-trimethylsilyl-3-butyn-2-yl succinate (**3**): ^1H NMR (300 MHz, $CDCl_3$) δ: 0.16 (s, 9 H), 1.47 (d, *J* = 6.6, 3 H), 2.50-2.80 (m, 4 H), 5.49 (q, *J* = 6.6, 1 H).

8. Physical characteristics of (*R*)-4-trimethylsilyl-3-butyn-2-ol (**4**): $[\alpha]_D^{20}$+22.4 (c = 2.01, $CHCl_3$); IR (film) cm^{-1} υ: 3334, 2175; ^{1}H NMR (300 MHz, $CDCl_3$) δ: 0.17 (s, 9 H), 1.45 (d, *J* = 6.6, 3 H), 1.83 (d, *J* = 5.3, 1 H), 4.52 (app p, *J* = 6.3, 1 H); ^{13}C NMR (75 MHz, $CDCl_3$) δ: -0.20, 24.1, 58.5, 88.2, 107.7.

9. (*S*)-4-Trimethylsilyl-3-butyn-2-ol (**1**): $[\alpha]_D^{20}$–22.3 (c = 2.55, $CHCl_3$); lit. (–25.9, c = 3.12).[2]

10. The acetate derivative of (*S*)-4-trimethylsilyl-3-butyn-2-ol was prepared by treatment with excess acetic anhydride, Et_3N, and DMAP in CH_2Cl_2. Analysis of an aliquot by GC on a β-Dex column (80°C, 0.2°C ramp/min) revealed a 98:2 mixture of enantiomers.

11. Physical characteristics of (*R*)-4-trimethylsilyl-3-butyn-2-yl mesylate (**5**): $[\alpha]_D^{20}$ +98.4 (c = 2.30, $CHCl_3$); IR (film) cm^{-1}: 2961, 2175; ^{1}H NMR (300 MHz, $CDCl_3$) δ: 0.15 (s, 9 H), 1.58 (d, *J* = 6.7, 3 H), 3.07 (s, 3 H), 5.20 (q, *J* = 6.7, 1 H); ^{13}C NMR (75 MHz, $CDCl_3$) δ: –0.63, 22.3, 38.9, 68.3, 93.4, 101.1.

12. Care must be taken not to heat the bath during removal of the solvent from the relatively volatile mesylate solution on the rotary evaporator.

13. Physical characteristics of (*R*)-3-butyn-2-ol mesylate (**6**): $[\alpha]_D^{20}$ +92.9 (c = 2.49, $CHCl_3$); IR (film) cm^{-1}: 3282, 2942, 2124; ^{1}H NMR (300 MHz, $CDCl_3$) δ: 1.63 (d, *J* = 6.6, 3H), 2.72 (d, *J* = 1.8, 1H), 3.10 (s, 3H), 5.27 (dq, *J* = 1.7, 6.4 Hz, 1H); ^{13}C NMR (75 MHz, $CDCl_3$) δ: 22.3, 39.0, 67.4, 76.3, 80.0.

14. Tributyltin hydride was purchased from Lancaster Synthesis, Inc.

15. Addition of the $CuBr \cdot SMe_2$ over 5 min in small portions is important to the success of the reaction.

16. After addition of the mesylate to the stirred reaction mixture, it is necessary to quench the reaction within 10 min to minimize copper-catalyzed racemization of the allenylstannane. In our original communication of this methodology[3] we noted that, contrary to the findings of others[4], who reported partial racemization of allenes prepared through additions of alkylcuprates to propargylic esters, the acetoxymethyl stannane (eq 5, R^1 = $AcOCH_2$, R^2 = Me) was not racemized by prolonged exposure to the stannylcupration conditions. It appears that, perhaps not surprisingly, structural factors can influence the configurational stability of allenes prepared in this manner. It is also possible that contaminants in the copper

salts or impurities of an undetermined nature could be responsible for the racemization, which is mechanistically obscure at this time.

17. Physical characteristics of (*P*)-(+)-3-(tributylstannyl)-1,2-butadiene (**7**): $[\alpha]_D^{20}$ +61.2 (c = 3.31, $CHCl_3$); IR (film) cm^{-1}: 2924, 1928; ^1H NMR (300 MHz, $CDCl_3$) δ: 0.80-1.70 (m, 30 H), 4.45-4.71 (m, 1 H), 4.87-5.12 (m, 1 H). ^{13}C NMR (75 MHz, $CDCl_3$) δ: 10.3, 13.7, 27.2, 28.9, 74.3, 75.2, 209.0. The ee of the stannane was determined by analysis of the product obtained through $BF_3 \cdot OEt_2$-promoted addition to isobutyraldehyde. To a stirred solution of (*P*)-(+)-3-(tributylstannyl)-1,2-butadiene (1.97 g, 57.5 mmol) and isobutyraldehyde (300 mg, 42 mmol) in CH_2Cl_2 at -78°C was added $BF_3 \cdot OEt_2$ (2.18 mL, 172 mmol). The solution was maintained at -78°C for 1 hr then quenched by the addition of sat. aq. $NaHCO_3$ and warmed to rt. The CH_2Cl_2 layer was removed and stirred vigorously with KF-on-Celite[5] and $MgSO_4$. After 1 hr, analysis of an aliquot by GC on a β-Dex column indicated a 98:2 ratio of enantiomers, and a >99:1 ratio of diastereomers (80°C, 0.2 °C ramp/min; alcohol 21.8 min).

Waste Disposal Information

All hazardous materials should be handled and disposed of in accordance with "Prudent Practices in the Laboratory"; National Academy Press; Washington, DC, 1995.

3. Discussion

The present resolution procedure is based on a report by Burgess and Jennings,[2] but with several key modifications.

1) To minimize losses of the volatile products during solvent removal, pentane was substituted for hexane.

2) For the same reason, and to facilitate larger scale resolutions, the substrate concentration was increased from 0.1 to ~0.2M. However, an increase to 0.4 M led to diminished efficiency.

3) In the original report, the (*S*)-alcohol **1** and (*R*)-acetate **2** were separated by column chromatography which led to isolated yields of 90% for the acetate and only 54% for the alcohol. In the present procedure the alcohol is separated from the acetate by conversion to the half succinate, which is extracted with aqueous sodium bicarbonate. Purification of the acetate is achieved by short-path distillation. After cleavage of the ester

groupings through reduction with DIBAL-H, both the (*S*) and (*R*)-alcohols are isolated following distillation in ~86% overall yield.

Me3Si—≡—C(OMs)(H)Me **ent-5**

R–CHO → Pd(OAc)2•PPh3, Et2Zn → **8**

R = c-C_6H_{11} (84%), iPr (86%), $PhCH_2CH_2$ (78%)
anti:syn 90:10 - 96:4; er 97:3 - 99:1

(1)

Me3Si—≡—C(OMs)(Me)H **5**

R^1–CHO → Pd(OAc)2•PPh3, InI → **9**

R^1 = c-C_6H_{11} (84%), iPr (86%), $PhCH_2CH_2$ (78%)
anti:syn 98:2 - 99:1; er 99:1

The (*R*)-TMS butynyl mesylate **5**, or its enantiomer, are converted *in situ* to the (*P*)-allenyl indium or zinc reagents by reaction with InI or Et_2Zn utilizing $Pd(OAc)_2 \cdot PPh_3$ as the precatalyst.[6] These metallations are conducted in the presence of various aldehydes to produce *anti* homopropargylic alcohol adducts **8** or **9** of high diastereomeric and enantiomeric purity (eq 1). Additions to chiral α and β-oxygenated aldehydes proceed with a high degree of reagent control to yield *syn*, *anti* and *anti*, *anti* stereotriads (eq 2, 3).[7] These additions can also be effected with the corresponding allenyl indium and zinc reagents derived from the terminal alkynyl mesylate **6**, or its enantiomer (eq 4).[8] However, those additions proceed with significantly lower diastereoselectivity, most notably with unbranched aldehydes.

Me–CH(OTBS)–CHO

Me3Si—≡—C(OMs)(Me)H; Pd(OAc)2•PPh3, InI (87%) → syn, anti; dr = 90:10

Me3Si—≡—C(OMs)(H)Me; Pd(OAc)2•PPh3, InI (74%) → anti, anti; dr = 99:1

(2)

(3)

anti, anti; dr = 99:1

syn, anti; dr = 99:1

Propargylic mesylates, such as **6**, can be converted to (*P*)-allenylstannanes through S_N2' displacement of the mesylate with a tributyltin cuprate reagent (eq 5).[9,10] These reagents are formed with inversion of stereochemistry and most are sufficiently stable to be purified by distillation. The reaction proceeds with high levels of enantioselectivity, provided the

(4)

R = c-C_6H_{11} (86%), iPr (71%), $PhCH_2CH_2$ (71%)
anti:syn 77:23 - 95:5; er 99:1

(5)

product is removed from the cuprate salts within 10 min. The allenyl tin reagents afford *syn* homopropargylic alcohol adducts upon addition to aldehydes in the presence of excess $BF_3{\cdot}OEt_2$ (eq 6). Matching/mismatching

(6)

R^1	R^2	Yield (%)	syn:anti
C_7H_{15}	C_6H_{13}	83	37:63
C_7H_{15}	t-Bu	92	99:1
C_7H_{15}	i-Pr	95	95:5
CH_2OAc	i-Pr	98	95:5

occurs in additions to α-chiral aldehydes (eq 7). Allenylstannanes undergo transmetallation with $BuSnCl_3$, $SnCl_4$, or InX_3 salts in the presence of aldehydes to afford allenylcarbinols or *anti* homopropargylic alcohols (eq 8).[11-13] The $SnCl_4$ transmetallations proceed with overall inversion of stereochemistry whereas the InX_3 reactions lead to allenylindium intermediates with retention of stereochemistry (eq 9).[14] The resulting allenyl $SnCl_3$ and InX_2 reagents afford enantiomeric *anti* homopropargylic

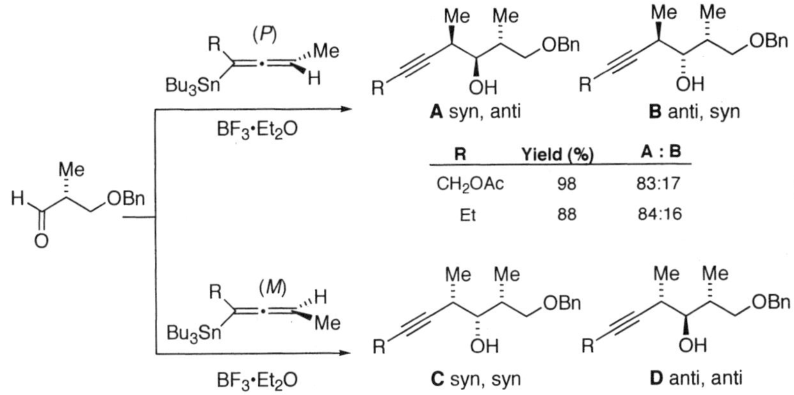

R	Yield (%)	A : B
CH_2OAc	98	83:17
Et	88	84:16

R	Yield (%)	C : D
CH_2OAc	96	>99:1
Et	92	>99:1

(7)

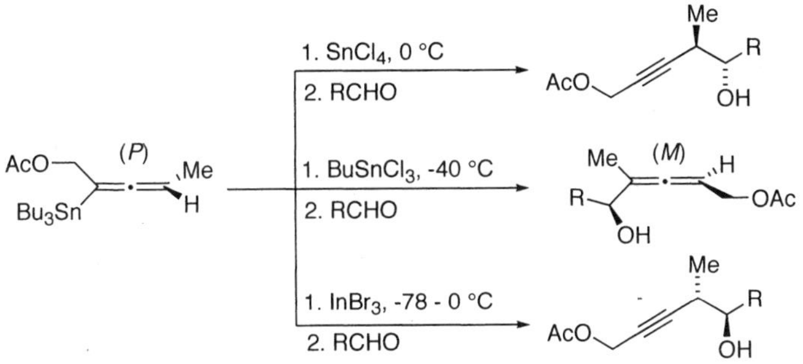

(8)

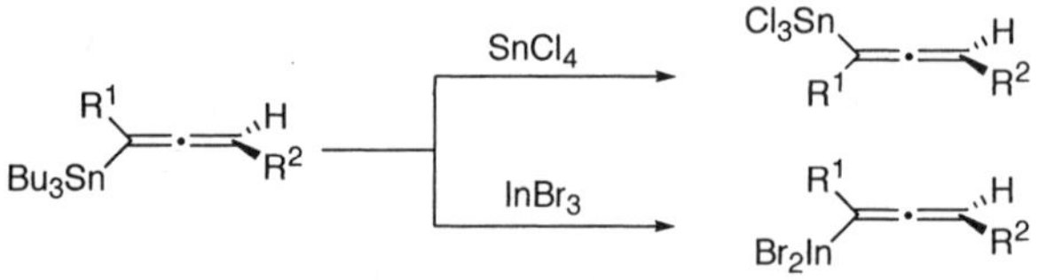

alcohols upon addition to aldehydes (eq 8). The $BuSnCl_3$ reaction proceeds via the transmetallated propargylic stannanes (eq 10). All three of the foregoing additions proceed through cyclic transition states.

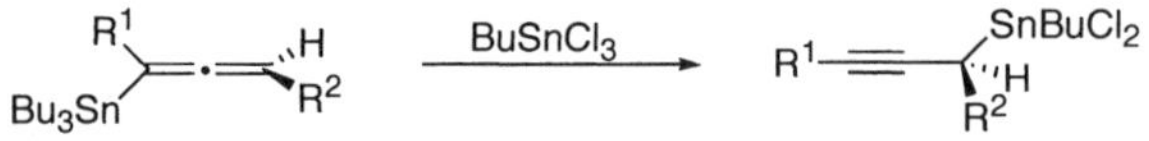

Additions to α-chiral aldehydes exhibit matching/mismatching characteristics (eq 11-13). Through appropriate selection of chiral partners and allenylmetal reagents, any of the eight possible stereotriads can be prepared by this methodology. The allenylcarbinol products are stereospecifically converted to dihydrofurans upon treatment with 5-10 mol% $AgNO_3$.[13]

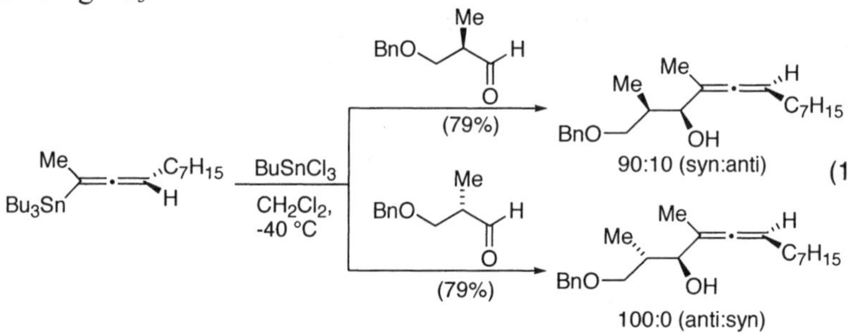

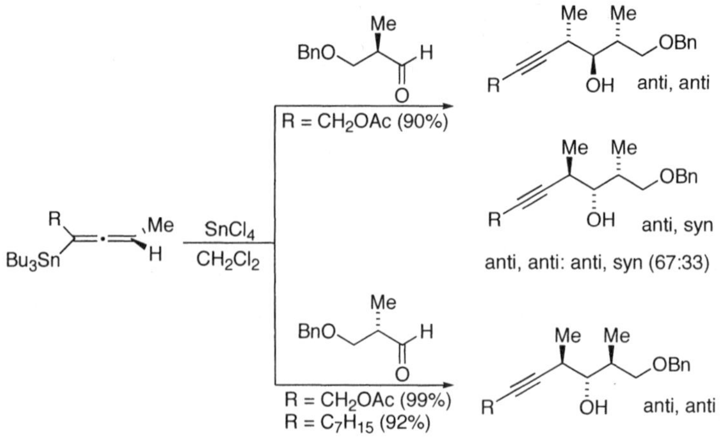

(13)

R = CH_2OAc, anti, syn:anti, anti = 95:5
R = H, anti, syn:anti, anti = 92:8

Chiral allenylmetal reagents have been utilized with excellent effectiveness in the synthesis of a variety of polyketide natural products of marine origin.[15, 16] An attractive feature of the methodology is the ability to incorporate the alkyne function of the adducts as a reactive component for elaboration of the polypropionate and polyacetate backbone of these compounds.

1. Department of Chemistry, McCormick Road, P.O. Box 400319, University of Virginia, Charlottesville, VA 22904.

2. Burgess, K.; Jennings, L. D. J. *Am. Chem. Soc.* **1991**, *113*, 6129.
3. Marshall, J. A.; Wang, X-J. *J. Org. Chem.* **1991**, *56*, 3211.
4. (a) LeQuan, M.; Cadiot, P. *Bull. Soc. Chem. Fr.* **1965**, 45. (b) LeQuan, M.; Guillem, G. J. *Organomet. Chem.* **1973**, *54*, 153. (c) Claesson, A.; Olsson, L. I. *J. Chem. Soc. Chem. Commun.* **1979**, 524.
5. Savall, B. M.; Powell, N. A.; Roush, W. R. *Org. Lett.* **2001**, *3*, 3057.
6. (a) Marshall, J. A.; Adams, N. D. *J. Org. Chem.* **1999**, *64*, 5201. (b) Marshall, J. A.; Grant, C. M. *J. Org. Chem.* **1999**, *64*, 8214.
7. Marshall, J. A.; Chobanian, H. R. *J. Org. Chem.* **2000**, *65*, 8357
8. Marshall, J. A.; Chobanian, H. R.; Yanik, M. M. *Org. Lett.* **2001**, *3*, 3369.
9. Marshall, J. A.; Wang, X-J. *J. Org. Chem.* **1990**, *55*, 6246.
10. Marshall, J. A.; Wang, X-J. *J. Org. Chem.* **1992**, *57*, 1242.
11. Marshall, J. A.; Perkins, J. *J. Org. Chem.* **1994**, *59*, 3509.
12. Marshall, J. A.; Perkins, J. F.; Wolf, M. A. *J. Org. Chem.* **1995**, *60*, 5556.
13. Marshall, J. A.; Yu, R. H.; Perkins, J. F. *J. Org. Chem.* **1995**, *60*, 5550.
14. Marshall, J. A.; Palovich, M. R. *J. Org. Chem.* **1997**, *62*, 6001.
15. (a) Marshall, J. A.; Johns, B. A. *J. Org. Chem.* **1998**, *63*, 7885. (b) Marshall, J. A.; Fitzgerald, R. N. *J. Org. Chem.* **1999**, *64*, 4477. (c) Marshall, J. A.; Johns, B. A. *J. Org. Chem.* **2000**, *65*, 1501. (d) Marshall, J. A.; Adams, N. D. *Org. Lett.* **2000**, *2*, 2897. (e) Marshall, J. A.; Yanik, M. M. *J. Org. Chem.* **2001**, *66*, 1373. (f) Marshall, J. A.; Schaaf, G. M. *J.*

Org. Chem. **2001**, *66*, 7825. (g) Marshall, J. A.; Adams, N. D. *J. Org. Chem.* **2002**, *67*, 733.

16. Several alternative non-resolution, large-scale routes to *(R)*- and *(S)*-3-butyn-2-ol have also been developed. The first two procedures require stoichiometric amounts of triphenylphosphine, whereas the latter employs lithiodichloromethane as the alkyne precursor; (a) Ku, Y.-Y.; Patel, R. R.; Elisseou, E. M.; Sawich, D. P. *Tetrahedron Lett.* **1995**, *36*, 2733. (b) Marshall, J. A.; Xie, S. *J. Org. Chem.* **1995,** *60*, 7230. (c) Marshall, J. A.; Yanik, M. M.; Adams, N. D.; Ellis, K. C.; Chobanian, H. R. *Org. Syn.* **2004**, *81*, 157.

Appendix
Chemical Abstracts Nomenclature; (Registry Number)

(*R*)-4-Trimethylsilyl-3-butyn-2-yl acetate: 3-Butyn-2-ol, 4-(trimethyl silyl)-, acetate; (129571-78-4)

(*S*)-4-Trimethylsilyl-3-butyn-2-yl succinate: Butanedioic acid, mono [(1*S*)-1-methyl-3-(trimethylsilyl)-2-propynyl] ester; (375395-73-6)

4-Trimethylsilyl-3-butyn-2-ol: 3-Butyn-2-ol, 4-(trimethylsilyl)-; (6999-19-5); (2*R*)-(121522-26-7); (2S)-(12155-27-8)

(*R*)-4-Trimethylsilyl-3-butyn-2-yl mesylate: 3-Butyn-2-ol, 4-(trimethylsilyl), methanesulfonate, (2*R*)-; (200440-90-0)

(*R*)-3-Butyn-2-yl mesylate: 3-Butyn-2-ol, methanesulfonate, (2*R*)-; (121882-95-4)

Vinyl acetate: Acetic acid ethenyl ester; (1008-05-4)

DIBAL-H (Diisobutylaluminum hydride): Aluminum, hydrobis(2-methylpropyl)-; (1191-15-7)

Triethylamine: Ethanamine, *N,N*-diethyl-; (121-44-8)

Mesyl chloride: Methanesulfonyl chloride; (124-63-0)

Butyllithium: Lithium, butyl-; (109-72-8)

Tributyltin hydride: Stannane, tributyl-; (688-73-3)

Cuprous bromide-dimethylsulfide complex: Copper, [thiobis(methane)]; (54678-23-8)

(P)-1-Tributylstannyl-1,2-butadiene: Stannane, 1,2-butadienyltributyl-; (202119-26-4)

IRIDIUM-CATALYZED SYNTHESIS OF VINYL ETHERS FROM ALCOHOLS AND VINYL ACETATE

(1-Methoxy-4-vinyloxybenzene)

H_3CO … OH + vinyl acetate → cat. $[Ir(cod)Cl]_2$, Na_2CO_3, toluene, 100 °C → H_3CO … O–vinyl

Submitted by Tomotaka Hirabayashi, Satoshi Sakaguchi, and Yasutaka Ishii.[1,2]
Checked by Paul W. Davies and Alois Fürstner.

1. Procedure

1-Methoxy-4-vinyloxybenzene. A 100-mL, two-necked round-bottomed flask is fitted with a magnetic stirbar, a reflux condenser connected to an argon/vacuum line, and a rubber septum. The equipment is flame dried under vacuum and then flushed with argon. Di-μ-chloro-bis(1,5-cyclooctadiene)diiridium(I), $[Ir(cod)Cl]_2$, (0.34 g, 0.01 eq., 0.5 mmol) (Note 1) and sodium carbonate (3.18 g, 30 mmol, 0.6 eq.) (Note 2) are rapidly weighed in air and added to the flask which is resealed, evacuated and backfilled with argon. Toluene (50 mL) (Note 3), *p*-methoxyphenol (6.21 g, 50 mmol) (Note 4) and vinyl acetate (8.61 g, 100 mmol) (Note 5) are successively introduced and the flask is placed into a preheated oil-bath at 100°C with magnetic stirring. The reaction mixture changes from a yellow color after the addition of the vinyl acetate to a wine red color within 30 minutes of heating (Note 6) and continues to darken over the course of the reaction. After heating for 2 hr, the reaction mixture is allowed to cool to ambient temperature before being transferred to a separatory funnel. Portions of EtOAc used to rinse the flask (3 x 50 mL) are added to the separatory funnel. The combined organic fractions are washed with water (3 x 100 mL) and brine (50 mL) before being dried over $MgSO_4$. The solution is filtered, the cake is washed with EtOAc (2 x 20 mL), and the combined filtrates are evaporated (Note 7). The resulting crude product is applied to a pre-packed silica gel column (50 x 3.5 cm) (Note 8) and eluted with hexanes/EtOAc (4:1) giving 1-methoxy-4-vinyloxybenzene as a pale yellow liquid after evaporation of the solvent (6.85 g, 91%) (Notes 7, 9, 10).

2. Notes

1. $[Ir(cod)Cl]_2$ can be prepared according to the literature method.[3] This complex is commercially available from Aldrich Chemical Company, Inc. The checkers used $[Ir(cod)Cl]_2$ purchased from Strem without further purification.

2. Na_2CO_3 purchased from Wako Pure Chemical Industries Ltd. was used as received. The checkers, however, found that the use of moist Na_2CO_3 led to lower yields of the desired product and increased amounts of the acetate side-product. Well reproducible results were obtained by using Na_2CO_3 dried under high vacuum (10^{-4} Torr) at 80°C overnight before use.

3. Toluene was dried by distillation over Na under Ar atmosphere.

4. *p*-Methoxyphenol was purchased from Wako Pure Chemical Industries Ltd. or Acros and used as received.

5. Vinyl acetate purchased from commercial suppliers (Wako Pure Chemical Industries Ltd., or Acros) was dried over MS 4Å and distilled at normal pressure (bp 72°C) under argon.

6. This characteristic color change is much slower if insufficiently dried Na_2CO_3 is used, thus leading to longer overall reaction times.

7. To avoid any loss of product during the evaporation of the solvent, the vacuum was set to ≥ 20 mbar and the temperature of the water bath should not exceed 40°C.

8. Silica gel (230-400 mesh) from Kanto Kagaku Reagent Division was used. The checkers used Merck silica gel (230-400 mesh).

9. A small amount (< 3%) of 1-acetoxy-4-methoxybenzene was observed as a side product. The checkers obtained somewhat larger amounts of this compound (0.52 g, 6%).

10. Analytical data: ^{1}H NMR (400 MHz, $CDCl_3$) δ: 3.79 (s, 3 H), 4.34 (dd, J = 1.7, 6.2 Hz, 1 H), 4.64 (dd, J = 1.7, 13.8 Hz, 1 H), 6.59 (dd, J = 6.2, 13.8 Hz, 1 H), 6.97-6.84 (m, 4 H); ^{13}C NMR (101 MHz, $CDCl_3$) δ: 55.8, 93.7, 114.8, 118.8, 149.6, 150.7, 155.8; IR (film) 2953, 2835, 1639, 1500, 1209, 1147, 1034, 954, 830 cm^{-1}. Anal. Calcd for $C_9H_{10}O_2$: C, 71.98; H 6.71; Found: C, 71.85; H, 6.67.

Safety and Waste Disposal Information

All hazardous materials should be handled and disposed of in accordance with "Prudent Practices in the Laboratory"; National Academy Press; Washington, DC, 1995.

3. Discussion

Vinyl ethers are conventionally prepared by mercury-catalyzed transfer vinylation,[4] base- or transition metal-catalyzed isomerization of allyl ethers,[5]

Table 1. Iridium-Catalyzed Synthesis of Various Vinyl Ethers

Entry	Alcohol	Product	Time	GLC Yield	Isolated Yield
1	4-bromophenol (Br, OH)	4-bromophenyl vinyl ether (Br, O)	3 h	91%	88%
2	4-chlorophenol (Cl, OH)	4-chlorophenyl vinyl ether (Cl, O)	3 h	92%	49%
3	4-isopropylphenol (OH)	4-isopropylphenyl vinyl ether (O)	3 h	98%	92%
4	H_3COCHN, OH	H_3COCHN, O	15 h	84%	79%
5[a]	HO, OH (biphenol)	O, O (divinyl ether)	6 h	88%	80%
6[a]	HO, OH (bisphenol)	O, O (divinyl ether)	6 h	90%	86%
7	1-phenylethanol (OH)	O	2 h	95%	-
8[b]	SH	S	24 h	92%	-
9[a]	HO–(CH2)10–OH (10)	O–(CH2)10–O (10)	6 h	95%	83%

[a] 4 eq. of vinyl acetate and 2 eq. of Na_2CO_3 were used. [b] See text.

or elimination reactions.[6] Quite recently, the synthesis of allyl and alkyl vinyl ethers by the transfer vinylation of alcohols with butyl vinyl ether in the presence of a palladium catalyst has been reported.[7]

The present method uses commercially available $[Ir(cod)Cl]_2$ as the precatalyst for the synthesis of vinyl ethers which are very difficult to prepare otherwise.[2] The reaction is thought to proceed through an addition-elimination sequence of alcohol and acetic acid in the presence of the iridium complex as a catalyst and Na_2CO_3 as the base. Representative examples are compiled in the Table. The method can also be applied to the synthesis of vinyl ethers derived from secondary alcohols (entry 7) and holds promise for the preparation of alkyl vinyl ethers (entry 9). Although sulfur compounds frequently inhibit transition metal-catalyzed reactions, thiophenol reacts with vinyl acetate to form phenylvinyl thioether in excellent yield (entry 8); this compound, however, is unstable in air and is therefore difficult to isolate in analytically pure form (~15 %).

1. Department of Applied Chemistry, Faculty of Engineering, Kansai University, Suita, Osaka 564-8680, Japan.
2. Okimoto, Y.; Sakaguchi, S.; Ishii, Y. *J. Am. Chem. Soc.* **2002**, *124*, 1590.
3. Herde, J. L.; Lambert, J. C.; Senoff, C. V. *Inorg. Synth.* **1974**, *15*, 18.
4. Watanabe, W. H.; Conlon, L. E. *J. Am. Chem. Soc.* **1956**, *79*, 2828.
5. Larock, R. C. *Comprehensive Organic Transformations,* 2nd ed.; Wiley-VCH: New York, **1999**, pp. 222 and 225-226.
6. Nerdel, F.; Buddrus, J.; Brodowski, W.; Hentschel, P.; Klamann, D., Weyerstahl, P. *Justus Liebigs Ann. Chem.* **1967**, *710*, 36.
7. Bosch, M.; Schlaf, M. *J. Org. Chem.* **2003**, *68*, 5225.

Appendix
Chemical Abstracts Nomenclature; (Registry Number)

Di-μ-chloro-bis(1,5-cyclooctadiene)diiridium(I): Iridium,

di-μ-chlorobis[(1,2,5,6-η)-1,5-cyclooctadiene]di-; (12112-67-3)

Sodium carbonate: Carbonic acid disodium salt; (497-19-8)

Toluene: Benzene, methyl-; (108-88-3)

p-Methoxyphenol: Phenol, 4-methoxy-; (150-76-5)

Vinyl acetate: Acetic acid ethenyl ester; (108-05-4)

1-(*tert*-BUTYLIMINOMETHYL)-1,3-DIMETHYLUREA HYDROCHLORIDE

[Urea, *N*-[[(1,1-dimethylethyl)imino]methyl]-*N,N'*-dimethyl-, monohydrochloride]

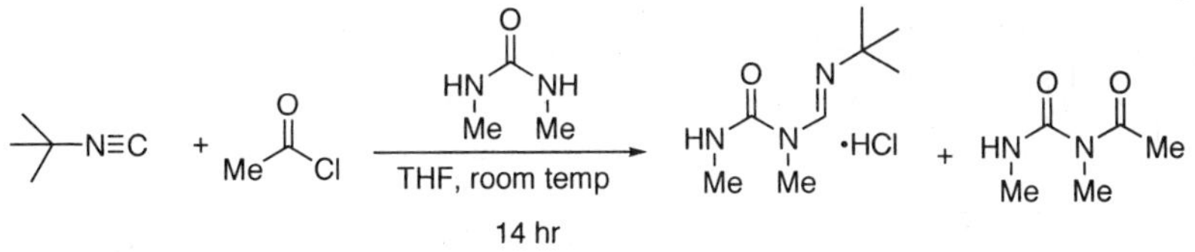

Submitted by David D. Díaz,[1] Amy S. Ripka,[1] and M.G. Finn.[1]
Checked by Daniel Laurich, Günter Seidel, and Alois Fürstner.

1. Procedure

Caution! Use safety glasses and nitrile gloves under a well-ventilated hood since isocyanides have pungent odors and some are known to be toxic.

1-(tert-Butyliminomethyl)-1,3-dimethylurea hydrochloride. A 1-L, three-necked, round-bottomed flask is equipped with a mechanical stirrer (Note 1), a vacuum take-off adapter attached to a nitrogen source, and a pressure-equalizing dropping funnel closed with a glass stopper. The flask is flame-dried, cooled under nitrogen, and charged with 250 mL of tetrahydrofuran (THF) (Note 2) and *tert*-butyl isocyanide (12.4 mL, 110 mmol) (Note 3). To this homogeneous solution is added acetyl chloride (8.6 mL, 121 mmol) (Note 4) and the mixture is vigorously stirred for 15 min. A solution of 1,3-dimethylurea (24.2 g, 275 mmol) (Note 3) in 150 mL of THF is then introduced slowly and continuously through the dropping funnel. After 15-20 min, a white precipitate appears (Note 5). The reaction mixture is stirred for 14 hr at room temperature, after which time the heterogeneous mixture is filtered through a Büchner funnel connected to vacuum. The white precipitate is washed with cold THF (Note 6) to remove excess urea, the acetylurea byproduct and colored impurities (Note 7). The product is dried in a vacuum oven at 50°C for 20 hr to give 19-20 g (85-90%) of analytically pure product as a white powder (Notes 8-10). The hygroscopic product should be stored under nitrogen.

2. Notes

1. The submitters used a magnetic stir bar (50 mm in length x 7 mm in diameter) instead of the mechanical stirrer, but noted that stirring may become difficult as the reaction proceeds. This problem is avoided when a mechanical stirrer is used.

2. The submitters used commercial tetrahydrofuran, grade OPTIMA (Fisher). The checkers used THF dried by distillation over Mg-anthracene under argon.

3. *tert*-Butyl isocyanide ($\geq$ 98%, Fluka) and 1,3-dimethylurea, 98% (Acros) were used as received.

4. Acetyl chloride, p.a., was used as supplied by Acros, and dispensed by syringe. For the scale proposed it is not necessary to distill this reagent, although this is recommended for smaller scale reactions.

5. The submitters noted that stirring becomes difficult shortly after the precipitation starts if a magnetic stir bar is used.

6. THF at 0°C is necessary for the washing process as follows: suction is stopped and the white solid is suspended completely in cold THF, the suction is turned on to remove the liquid, and the operation is repeated several times until 1 L of cold THF is used. After washing, air is pulled through the solid for 20 min to dry the product, and clumps are broken manually with a spatula. The solid remaining on the walls of the funnel is dissolved in CH_2Cl_2 and the solvent evaporated under vacuum. The two fractions of solid are combined.

7. The byproduct *N*-acylurea may be isolated in equimolar amounts to formamidine urea by column chromatography on silica gel (7:3 EtOAc:hexanes, R_f ca. 0.5) of the residue from the combined filtrates.

8. The checkers dried the product *in vacuo* (10^{-3} Torr) at ambient temperature for 3 hr.

9. Spectral and analytical data: mp 172-173°C; ^{1}H NMR (400 MHz, $CDCl_3$) δ: 1.59 (s, 9 H), 2.94 (d, J = 4.4 Hz, 3 H), 3.85 (s, 3 H), [8.72 (d, J = 15.2 Hz), 8.7 (m); 2 H], 11.09 (d, J = 14.8 Hz, 1 H); ^{13}C NMR (100 MHz, $CDCl_3$) δ: 28.0, 29.1, 34.7, 58.7, 150.7, 152.8; IR (film): 3209, 2984, 1722, 1671, 1535 cm^{-1}. Anal. calcd for $C_8H_{18}ClN_3O$: C, 46.26; H, 8.74; Cl, 17.09; N, 20.23. Found: C, 46.14; H, 8.63; Cl, 17.10; N, 20.29.

10. As a precaution, the formamidine urea should be handled with care in a well-ventilated hood.

Safety and Waste Disposal Information

All hazardous materials should be handled and disposed of in accordance with "Prudent Practices in the Laboratory"; National Academy Press; Washington, DC, 1995.

3. Discussion

Formamidines are of interest in synthetic chemistry[2] and have been used extensively as pesticides. The present procedure is an example of a novel process in which the addition of a substituted urea to a mixture of isocyanide and acid chloride gives formamidine urea salts **1** in pure form.[3] The yields of **1** are maximized by the use of 2.5-3.0 equiv of urea, and THF provides the best results in general. The process was found to be tolerant of substantial variations in the nature of the components, but restricted almost exclusively to acid chlorides. A proposed mechanism for the formation of **1** is shown in Scheme 1.[3] The first key intermediate is the electrophilic adduct of isonitrile and acid chloride, **2**. Substitution of chloride by urea at the chloroiminium carbon gives **3**. The unique aspect of the mechanism is then

Scheme 1

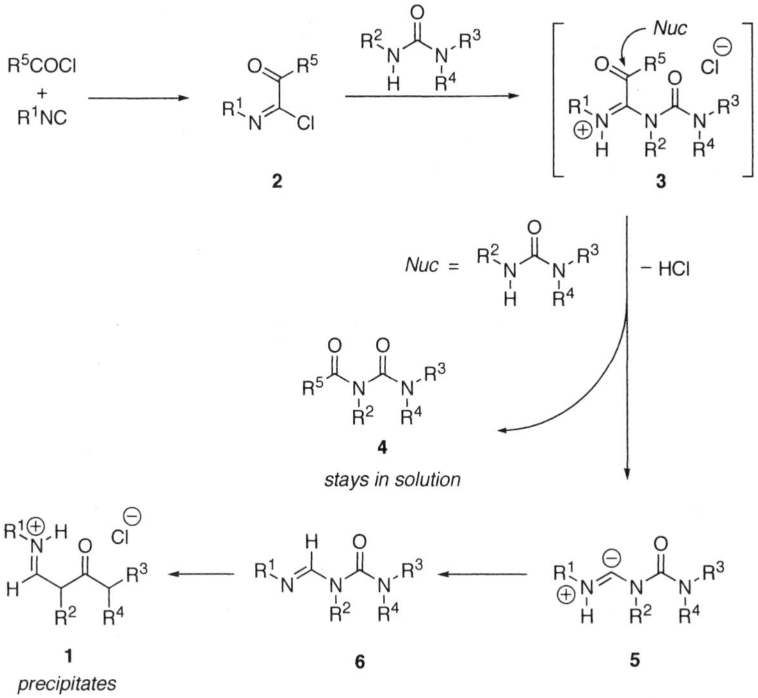

R^5COCl + R^1NC → **2** → **3** → (− HCl) **5** → **6** → **1**

Nuc = urea; **4** stays in solution; **1** precipitates

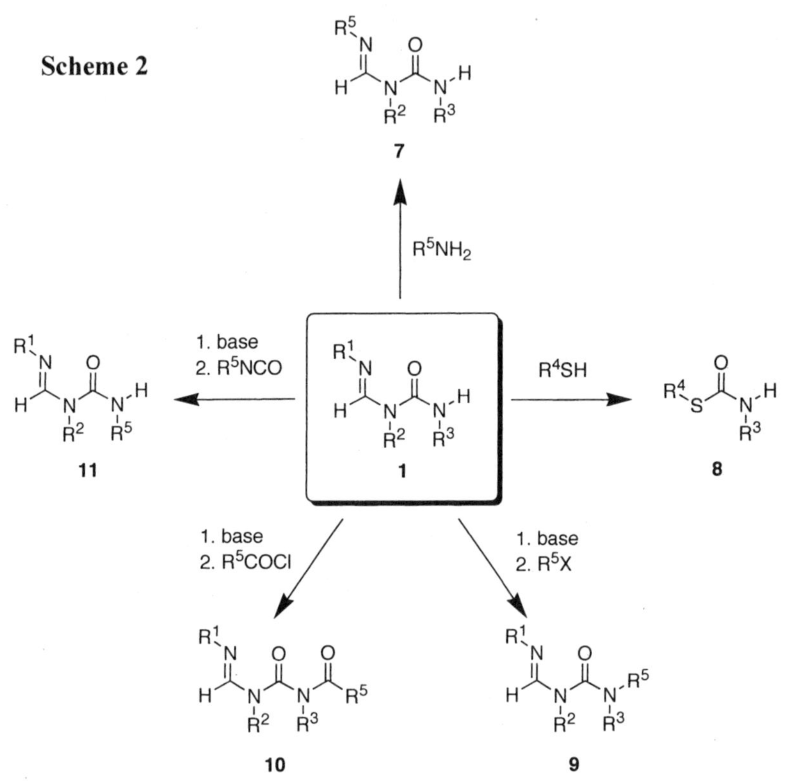

proposed to be capture of **3** at the active carbonyl carbon by another equivalent of urea to form the soluble acylurea **4** and intermediate **5**, which may be formulated as the betaine shown or a stabilized carbenoid. Structure **5** should undergo rapid proton transfer to give the formamidine **6**, and an equivalent of HCl is extracted by precipitation of the hydrochloride salt **1**. The efficiency and convenience of the reaction is governed by both the generation of the reactive adducts and the precipitation of the final product. The overall transformation is made possible by the ability of the acid chloride to change the nature of the isonitrile carbon from nucleophilic to electrophilic. The entry to the formamidine urea skeleton exemplified here is quite a bit more convenient than previous methods.

The *tert*-butylamino fragment of **1** may be readily exchanged with a wide variety of primary nitrogen nucleophiles,[4] allowing the easy preparation of a variety of formamidine ureas (**7**, Scheme 2). Formamidine ureas also undergo substitution at the carbonyl carbon to give thiolcarbamates (**8**) and the formamidine urea nucleus can be elaborated by base-mediated alkylation and acylation (**9**, **10**, **11**).[5]

1. Department of Chemistry and The Skaggs Institute for Chemical Biology, The Scripps Research Institute, 10550 N. Torrey Pines Rd., La Jolla, CA, 92037, USA. We thank Dr. Chunmei Li for her valuable assistance.
2. (a) Meyers, A. I.; Hutchings, R. H., *Tetrahedron* **1993**, *49*, 1807-1820. (b) Meyers, A. I.; Elworthy, T. R. *J. Org. Chem.* **1992**, *57*, 4732-4740, and references cited therein.
3. Ripka, A. S.; Díaz, D. D.; Sharpless, K. B.; Finn, M. G. *Org. Lett.* **2003**, *5*, 1531-1533.
4. Díaz, D. D.; Finn, M. G. *Chem. Eur. J.* **2004**, *10*, 303-309.
5. Díaz, D. D.; Finn, M. G. *Org. Lett.* **2004**, *6*, 43-46.

Appendix
Chemical Abstracts Nomenclature; (Registry Number)

1,3-Dimethylurea: Urea, *N,N'*-dimethyl-; (96-31-1)

Acetyl chloride; (75-36-5)

tert-Butyl isocyanide: Propane, 2-isocyano-2-methyl-; (7188-38-7)

ORTHO-FORMYLATION OF PHENOLS: PREPARATION OF 3-BROMOSALICYLALDEHYDE

(3-Bromo-2-hydroxybenzaldehyde)

$MgCl_2$, Et_3N, $(CH_2O)_n$ / THF, 2 hr

Submitted by Trond Vidar Hansen[1] and Lars Skattebøl.[2]
Checked by David Guthrie and Dennis P. Curran.

1. Procedure

3-Bromo-2-hydroxybenadehyde. A dry 500-mL, three-necked round-bottomed flask equipped with a stirring bar, reflux condenser and rubber septa is purged with argon gas. Anhydrous magnesium dichloride (Notes 1 and 2) (9.52 g, 100 mmol) and solid paraformaldehyde (Notes 3 and 4) (4.50 g, 150 mmol) are added, while a slight positive pressure of argon is maintained. Dry tetrahydrofuran (250 mL) (Note 5) is added by syringe. Triethylamine (Notes 6 and 7) (10.12 g, 100 mmol) is added drop-wise by syringe and the mixture is stirred for 10 min. 2-Bromophenol (Note 8) (8.65 g, 50 mmol) is added drop-wise by syringe, resulting in an opaque, light pink mixture. This mixture is immersed in an oil bath at about 75°C (bath temperature) and soon turns a bright orange-yellow color. Heating at gentle reflux is maintained for 4 hr (Note 9).

The reaction mixture is cooled to room temperature and 100 mL of ether is added. The resulting organic phase is transferred to a 1-L separatory funnel and washed successively with 1N HCl (3 x 100 mL) (*Caution! gas evolves*) and water (3 x 100 mL) (Note 10), dried over anhydrous magnesium sulfate ($MgSO_4$), and filtered. The solvent is removed by rotary evaporation leaving a pale yellow oil that solidifies on further vacuum drying at 1-2 mmHg. The resulting yellow solid (8.09-8.11 g, 80-81%) consisting mainly (≥95%) of 3-bromosalicylaldehyde (Note 11) is sufficiently pure for further synthetic use. Recrystallization from hexane (50 mL) gives 6.80-6.94 g (68-69%) (Note 12) of pure aldehyde as pale yellow needles (Note 13).

2. Notes

1. Magnesium dichloride, anhydrous beads, –10 mesh, 99.9% was purchased from the Aldrich Chemical Company, Inc. The submitters dried the beads over phosphorus pentoxide under reduced pressure for 24 hr prior to use. The checkers purchased the beads in ampules with less than 100 ppm water and used these directly. The use of anhydrous beads is crucial; the checkers observed little or no reaction when using anhydrous magnesium chloride powder dried over phosphorous pentoxide.

2. The submitters report that the reaction proceeds at a slower rate with less than two equivalents of magnesium dichloride and gives a lower yield of aldehyde.

3. Paraformaldehyde powder was purchased from Aldrich Chemical Company, Inc. and dried over phosphorus pentoxide under reduced pressure for 24 hr prior to use.

4. The reaction requires two equivalents of paraformaldehyde; however, the submitters report that an excess of this reagent results in a faster reaction and higher yield of aldehyde.

5. Tetrahydrofuran was purchased from Aldrich Chemical Company, Inc. The submitters distilled the THF from sodium benzophenone ketyl prior to use, while the checkers passed the THF through a column of dry, activated aluminum oxide.

6. Triethylamine was purchased from Aldrich Chemical Company, Inc., distilled from calcium hydride and stored over 4Å molecular sieves prior to use.

7. Equimolar amounts of triethylamine and magnesium dichloride are required.

8. 2-Bromophenol was purchased from Aldrich Chemical Company, Inc. and used as received.

9. Progress of the reaction was followed by thin layer chromatography (TLC) using Merck silica gel 60 F254 aluminum-backed plates, eluting with hexane/EtOAc (10:1) and visualizing with a 254 nm UV lamp: 2-bromophenol $R_f \approx 0.58$; 3-bromsalicylaldehyde $R_f \approx 0.45$.

10. Vigorous shaking of the separatory funnel should be avoided since emulsions can result. The mixture is swirled and tilted back and forth gently for 3-4 min per washing. This results in rapid separation of the layers on standing.

11. The main impurity is the starting bromophenol. The product can be analyzed by GLC on a Varian GC 3300 instrument equipped with a 25 m SP2100 capillary column or by ^1H NMR spectroscopy.

12. The submitters obtained 9.05 g (90%) of the recrystallized product.

13. 3-Bromosalicylaldehyde exhibited the following physical and spectral properties: mp (uncorr.) 52-53°C (lit.[3] 52°C); ^{1}H NMR (300 MHz, $CDCl_3$) δ: 6.94 (t, J = 7.8 Hz), 7.55 (dd, J = 1.3 Hz, 7.7 Hz), 7.75 (dd, J = 1.3 Hz, 7.7 Hz), 9.85 (s, 1 H), 11.60 (s, 1 H); ^{13}C NMR (75 MHz, $CDCl_3$) δ: 111.0, 120.7, 121.2, 132.9, 139.9, 157.9, 196.0; FTIR spectrum ($CHCl_3$) cm^{-1}: 3152, 2852, 1662, 906; LRMS (EI) *m/z* 202/200 (M^+), 159, 149, 117, 115, 71, 57, 55; HRMS Calcd. for $C_7H_5BrO_2$: 199.9473; Found: 199.9475.

Safety and Waste Disposal Information

All hazardous materials should be handled and disposed of in accordance with "Prudent Practices in the Laboratory"; National Academy Press; Washington, DC, 1995.

3. Discussion

Surprisingly few substituted salicylaldehydes are commercially available even though they are important intermediates in organic synthesis for the preparation of a variety of oxygen-containing heterocyclic compounds and as a source for salen ligands.[5] Salicylaldehydes are accessible from the corresponding phenols by several classical formylation reactions;[4] however, the yields are often moderate and the lack of regioselectivity is problematic.

The present procedure is a simple, efficient and regioselective method for the preparation of substituted salicylaldehydes that is based on the work reported by Hofsløkken and Skattebøl.[6] It is applicable for large-scale preparations; on a 0.25 molar scale, 3-bromosalicylaldehyde was obtained in 90% yield. The method gives exclusively *ortho*-formylation of phenols and naphthols; no bis-formylation has been observed. Tetrahydrofuran may be replaced by acetonitrile with a negligible effect on yields. Reactions usually require 2-4 hours for completion. Electron releasing substituents enhance the reaction rate while the opposite effect is observed for electron withdrawing substituents. Prolonged reaction times increase the amounts of byproducts, particularly formation of the corresponding 2-methoxymethylphenol derivatives.[6]

Some representative examples of salicylaldehydes prepared by this method are compiled in the Table.

TABLE
ORTHO FORMYLATION OF PHENOLS AND NAPHTHOLS WITH $MgCl_2$/PARAFORMALDEHYDE

Phenol	Salicylaldehyde	Yield[Ref]
OH, F	OH, F, CHO	80%[7]
OH, Cl	OH, Cl, CHO	87%[6]
OH, Me	OH, Me, CHO	99%[6]
OH, F	OH, CHO, F	85%[7]
OH, Br	OH, CHO, Br	91%[7]
OH, OMe	OH, CHO, OMe	87%[6]
OH, CO_2Me	OH, CHO, CO_2Me	88%[6]
OH, *t*-Bu, *t*-Bu	OH, *t*-Bu, CHO, *t*-Bu	84%[7]
OH, Me	OH, CHO, Me	90%[6]
OH	CHO, OH	70%[6]

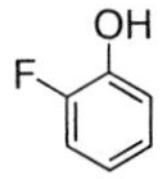

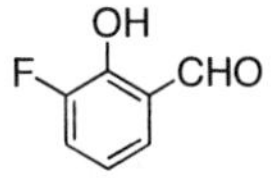

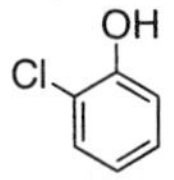

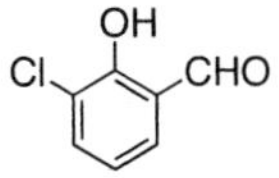

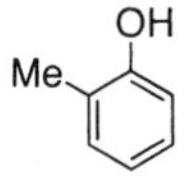

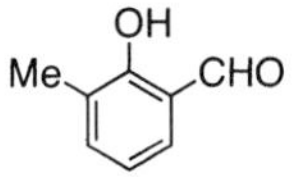

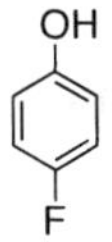

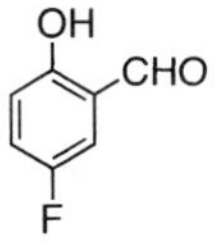

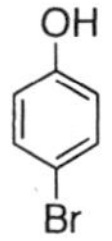

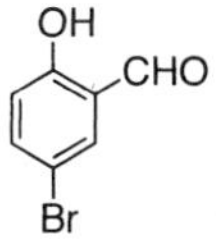

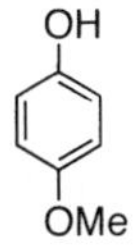

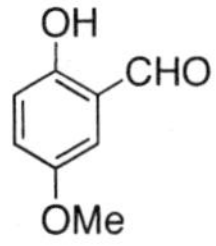

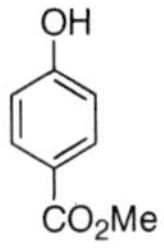

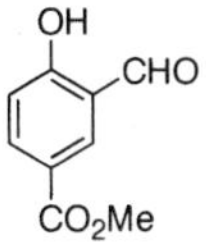

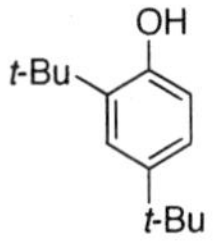

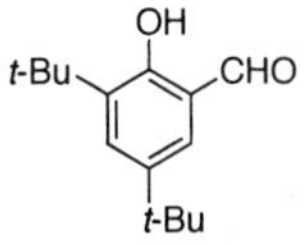

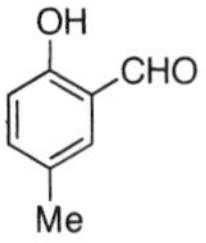

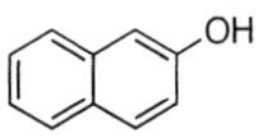

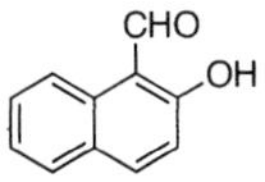

1. Department of Medicinal Chemistry, School of Pharmacy, University of Oslo, POB 1185, Blindern, N-0316 Oslo, Norway.
2. Department of Chemistry, University of Oslo, POB 1033, Blindern, N-0315 Oslo, Norway.

3. Müller, J. *Ber.* **1909**, *42*, 1166.
4. For a review, see: Laird, T. in *Comprehensive Organic Chemistry*, Stoddart, J. F., Ed., Pergamon, Oxford 1979, Vol. 1, p 1105.
5. Larrow, J. F.; Jacobsen, E. N.; Gao, Y.; Hong, Y., Nie; X. Zepp, C. M. *J. Org. Chem.* **1994**, *59*, 1939.
6. Hofsløkken, N. U.; Skattebøl, L. *Acta Chem. Scand.* **1999**, *53*, 258.
7. Hansen, T. V.; Skattebøl, L., *unpublished work.*

Appendix
Chemical Abstracts Nomenclature (Collective Index Number); (Registry Number)

Triethylamine: Ethanamine, *N,N*-diethyl- (9); (121-44-8)

Paraformaldehyde (9); (30525-89-4)

2-Bromophenol: Benzene, 2-bromo-1-hydroxy- (9); (95-56-7)

Magnesium chloride (8); (7786-30-3)

3-Bromosalicylaldehyde: Benzaldehyde, 2-hydroxy-3-bromo- (9); (1829-34-1)

PREPARATION OF 1-METHOXY-2-(4-METHOXYPHENOXY)BENZENE

[Benzene, 1-methoxy-2-(4-methoxyphenoxy)-]

Br OMe + HO OMe $\xrightarrow[\text{CuCl}]{Cs_2CO_3}$ O OMe OMe

Submitted by Elizabeth Buck and Zhiguo J. Song.[1]
Checked by Scott E. Denmark and John D. Baird.

1. Procedure

1-Methoxy-2-(4-methoxyphenoxy)benzene. To a 500-mL, three-necked, round-bottomed flask equipped with reflux condenser, nitrogen inlet, thermocouple and mechanical stirrer is added 14.0 g (113 mmol, 2.0 equiv) of 4-methoxyphenol (Note 1) and 95 mL of 1-methyl-2-pyrrolidinone (NMP) (Note 2). To this solution is then added 36.8 g (113 mmol, 2.0 equiv) of cesium carbonate (Note 3). The resulting slurry is degassed by evacuation and filling with nitrogen three times. Then 10.5 g (56.1 mmol) of 2-bromoanisole (Note 4) and 1.0 g (5.4 mmol, 0.096 equiv)) of 2,2,6,6-tetramethylheptane-3,5-dione (Note 5) are added followed by 2.8 g (28 mmol, 0.5 equiv) of copper (I) chloride (Note 6). The reaction mixture is degassed by evacuation and filling with nitrogen three times and then is heated to 120°C under nitrogen. The reaction is monitored by HPLC analysis (Note 8) and is stopped after 25 h when >98% conversion (by peak area) is obtained. The reaction mixture is cooled to room temperature and is diluted with 125 mL of methyl *t*-butyl ether (MTBE). The slurry is filtered and filter cake is washed with 125 mL of MTBE. The combined filtrates are washed subsequently with 175 mL of aqueous 2N HCl solution, 175 mL of 0.6N aqueous HCl solution, 150 mL of 2M aqueous NaOH solution, and 150 mL of 10% aqueous NaCl solution. The resulting organic layer is dried over $MgSO_4$ and is concentrated by rotary evaporation under vacuum at $\leq$ 30°C to remove all solvents. The crude, brown solid is triturated by magnetic stirring in 55 mL of hexanes (Note 9), then is aged for 4 hr and filtered to give 9.9-10.1 g (78% average yield) of an off-white solid (mp 77-78°C) of 99% purity as determined by HPLC analysis (Note 10).

2. Notes

1. 4-Methoxyphenol was purchased from Aldrich Chemical Co. and was used as received. Use of less than two equivalents resulted in a longer reaction time.

2. Anhydrous 1-methyl-2-pyrrolidinone (NMP) was purchased from Aldrich Chemical Co. and was used as received. NMP that was dried over 4Å molecular sieves with water content $\leq$100 μg/mL (Karl Fisher titration) gave slightly faster reactions.

3. Cesium carbonate was purchased from Alfa Aesar and was used as received. It is hygroscopic and should be weighed with minimum exposure to the air. Use of less than two equivalents resulted in slower reactions.

4. 2-Bromoanisole was purchased from Aldrich Chemical Co. and was used as received.

5. 2,2,6,6-Tetramethylheptane-3,5-dione was purchased from Aldrich Chemical Co. and was used as received.

6. Copper (I) chloride was purchased from Acros and was used as received.

7. MTBE (methyl t-butyl ether) was purchased from EM Science and was used as received.

8. The HPLC instrument used was a HP1100. HPLC conditions: Perkin Elmer 3X3 CR C18; MeCN/0.1% H_3PO_4; 2 mL/min; gradient: time 0 min 25/75, 1 min 25/75, 9 min 95/5, 10 min 95/5; UV detector 210 nm. Retention times: 4-methoxyphenol 0.6 min, 2-bromoanisole 3.4 min, product 4.3 min.

9. Hexanes was purchased from Fisher Scientific Co. and used directly as received.

10. Purity is reported as HPLC area percent (A%). Assay yields are reported using isolated material as standard. Analytical Data: ^{1}H NMR ($CDCl_3$, 400 MHz) δ: 7.09-7.05 (m, 1 H), 7.00-6.98 (m, 1 H), 6.97-6.93 (m, 2 H), 6.90-6.84 (m, 4 H), 3.88 (s, 3 H), 3.79 (s, 3H); ^{13}C NMR ($CDCl_3$, 100 MHz) δ: 155.3, 150.8, 150.7, 146.6, 123.7, 120.9, 119.3, 119.0, 114.6, 112.5, 55.9, 55.6. Anal. Calcd. for $C_{14}H_{14}O_3$: C, 73.03; H, 6.13. Found: C, 73.27; H, 6.05. MS (EI) *m/z* 230 (M^+, 100).

Safety and Waste Disposal Information

All hazardous materials should be handled and disposed of in accordance with "Prudent Practices in the Laboratory"; National Academy Press; Washington, DC, 1995.

3. Discussion

The Ullmann ether formation has traditionally been carried out under rather harsh conditions, usually at high temperatures in pyridine as the solvent.[1-3] The yields are typically low to moderate and reactions between electron-rich aryl halides and electron-deficient phenols typically do not work well. Other milder conditions, employing more expensive reagents or catalysts such as palladium or copper (II) triflate, are also available.[5,6] The major advantages of the process reported here are low cost of the copper additive and the simplicity of operation. Relative to other less expensive ligands for copper such as pyridine-based ligands or phosphine-based ligands,[5a,b] 2,2,6,6-tetramethylheptane-3,5-dione (TMHD) accelerates the Ullmann coupling reaction more effectively, allowing some of the electronically unfavorable substrates to undergo ether formation readily, as listed in Table 1. Nevertheless, these reactions exhibit the general trend that electron-donating groups on the phenol and electron-withdrawing groups on the halide make the reaction more favorable. Some tolerance of electron-withdrawing groups on the phenols was observed. The major side reaction in these Ullmann ether formations is the reduction of the aryl halides to the arene (Aryl-Br to Aryl-H). The reaction of an aryl iodide is faster and higher yielding than the corresponding bromide, as indicated by a comparison of entries 12 and 1.

Table 1. 2,2,6,6-Tetramethylheptane-3,5-dione-Catalyzed Ullmann Ether Formation

Ar-X	ArOH	ether product	reaction time (h)	assay yield (%)	isolated yield (%)
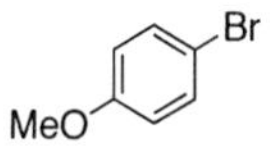 MeO, Br	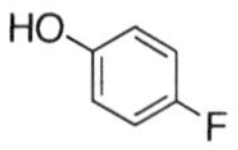 HO, F	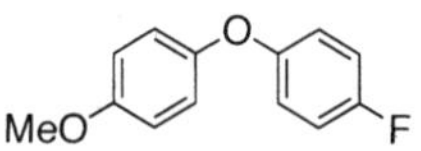 MeO, O, F	10	70	55
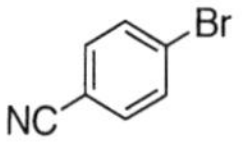 NC, Br	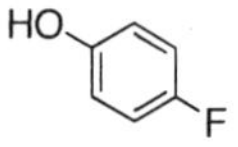 HO, F	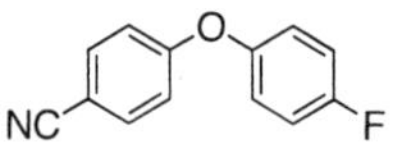 NC, O, F	1.5	90	82
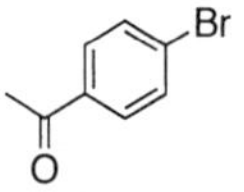 O, Br	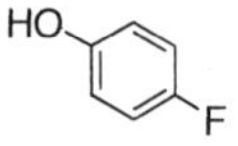 HO, F	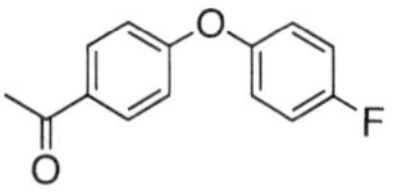 O, O, F	4	83	60
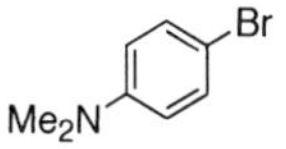 Me_2N, Br	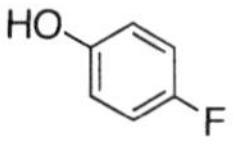 HO, F	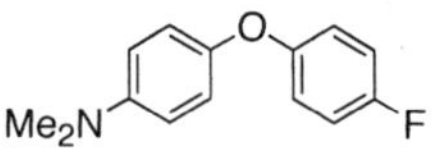 Me_2N, O, F	15	79	59
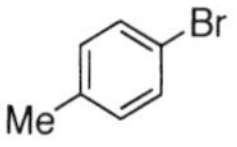 Me, Br	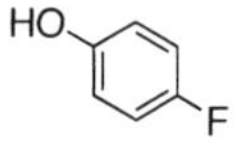 HO, F	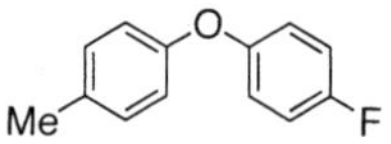 Me, O, F	7.5	86	63
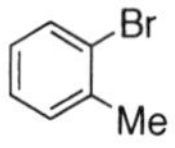 Br, Me	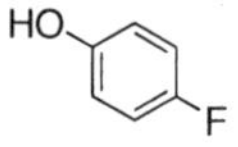 HO, F	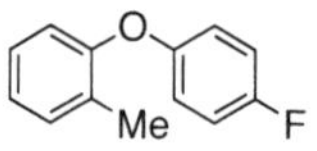 O, Me, F	15	97	85
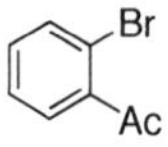 Br, Ac	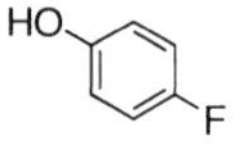 HO, F	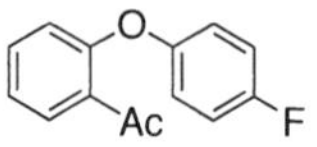 O, Ac, F	24	67	61
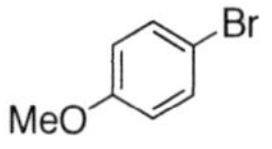 MeO, Br	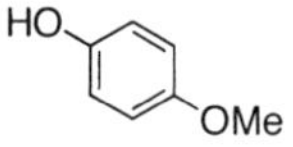 HO, OMe	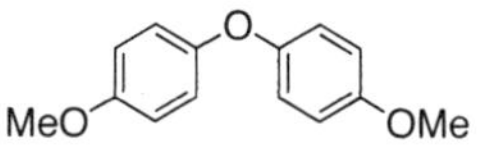 MeO, O, OMe	10	85	77
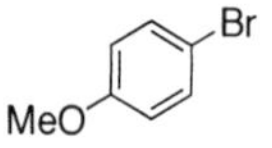 MeO, Br	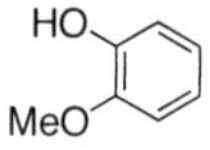 HO, MeO	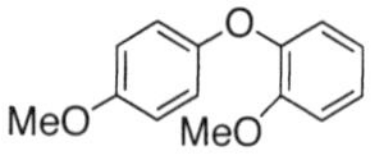 MeO, O, MeO	10	80	66
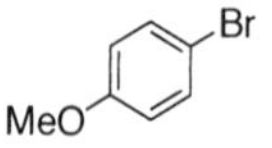 MeO, Br	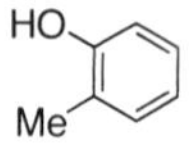 HO, Me	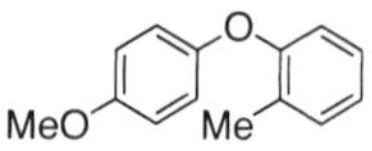 MeO, O, Me	10	92	79
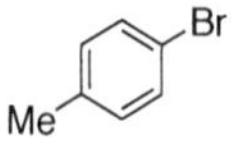 Me, Br	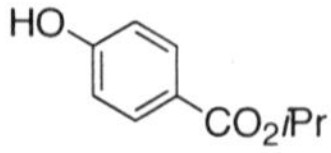 HO, CO_2iPr	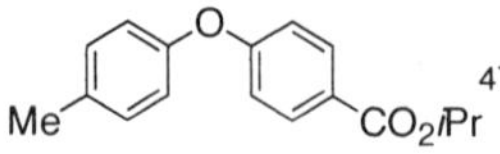 Me, O, CO_2iPr	47	56	51
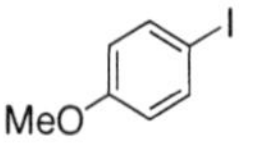 MeO, I	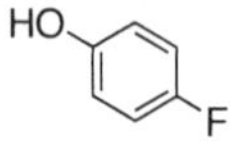 HO, F	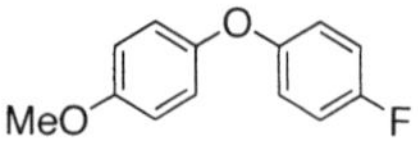 MeO, O, F	6	77	68

1. Department of Process Research, Merck Research Laboratories, P.O. Box 2000, Rahway, NJ 07065.
2. This procedure was first reported by: Buck, E.; Song, Z. J.; Tschaen, D.; Dormer, P. G.; Volante, R. P.; Reider, P. J. Org. Lett. 2002, 4, 1623.
3. For reviews: (a) Sawyer, J. S. Tetrahedron 2000, 56, 5045. (b) Theil, F. Angew Chem., Int. Ed. 1999, 38, 2345. (c) Lindley, J. Tetrahedron 1984, 40, 1433.
4. Some earlier reports: (a) Williams, A. L.; Kinney, R. E.; Bridger, R. F. *J. Org. Chem.* **1967**, *32,* 2501. (b) Bacon, R. G. R.; Rennison, S. C. *J. Chem. Soc.(C)* **1969**, 312. (c) Bacon, R. G. R.; Stewart, O. J. *J. Chem. Soc.* **1965**, 4953.
5. For palladium mediated ether synthesis: (a) Torraca, K. E.; Huang, X.; Parrish, C. A.; Buchwald, S. L. *J. Am. Chem. Soc.* **2001**, *123*, 10770. (b) Aranyos, A.; Old, D. W.; Kiyomori, A.; Wolfe, J. P.; Sadighi, J. P.; Buchwald, S. L. *J. Am. Chem. Soc.* **1999**, *121*, 4369. (c) Mann, G.; Hartwig, J. F. *Tetrahedron Lett.* **1997**, *38*, 8005. (d) Mann, G.; Incarvito, C.; Rheingold, A. L.; Hartwig, J. F. *J. Am. Chem. Soc.* **1999**, *121*, 3224. (e) Shelby, Q.; Kataoka, N.; Mann, G.; Hartwig, J. *J. Am. Chem. Soc.* **2000**, *122*, 10718.
6. For some recent reports of copper mediated ether formations: (a) Gujadhur, R. K.; Bates, C. G.; Venkataraman, D. *Org. Lett.* **2001**, *3,* 4315. (b) Fagan, P. J.; Hauptman, E.; Shapiro, R.; Casalnuovo, A. *J. Am. Chem. Soc.* **2000**, *122*, 5043. (c) Marcoux, J.-F.; Doye, S.; Buchwald, S. L. *J. Am. Chem. Soc.* **1997**, *119*, 10539. (d) Kalinin, A. V.; Bower, J. F.; Riebel, P.; Snieckus, V. *J. Org. Chem.* **1999**, *64*, 2986. (e) Evans, D. A.; Katz, J. L.; West, T. R. *Tetrahedron Lett.* **1998**, *39,* 2937. (f) Palomo, C.; Oiarbide, M.; Lopez, R.; Gomez-Bengoa, E. *Chem. Commun.* **1998**, 2091. (g) Smith, K.; Jones, D. *J. Chem. Soc. Perkin Trans. I* **1992**, 407. (g) Wolter, M.; Nordmann, G.; Job, G. E.; Buchwald, S. L. *Org. Lett.* **2002**, *4,* 973.

Appendix
Chemical Abstracts Nomenclature

4-Methoxyphenol: Phenol, 4-methoxy-; (150-76-5)

1-Methyl-2-pyrrolidinone (NMP): 2-Pyrrolidinone, 1-methyl-; (872-50-4)

Cesium carbonate: Carbonic acid, dicesium salt; (534-17-8)

2-Bromoanisole: Benzene, 1-bromo-2-methoxy-; (578-57-4)

2,2,6,6-Tetramethylheptane-3,5-dione: 3,5-Heptanedione, 2,2,6,6-tetramethyl-: (1118-71-4)

Copper (I) chloride: Copper chloride: (7758-89-6)

Methyl *t*-butyl ether (MTBE): Propane, 2-methoxy-2-methyl-: (1634-04-4)

D-RIBONOLACTONE AND 2,3-ISOPROPYLIDENE-(D-RIBONOLACTONE)

(D-Ribonic acid, 2,3-*O*-(1-methylethylidene)-, γ-lactone)

A. D-ribose → D-ribonolactone (Br_2, $NaHCO_3$, H_2O)

B. D-ribonolactone → 2,3-isopropylidene-D-ribonolactone (2,2-dimethoxypropane; Ag_2CO_3, H_2SO_4, acetone)

Submitted by John D. Williams,[1] Vivekanand P. Kamath,[2] Philip E. Morris,[2] and Leroy B. Townsend.[1]
Checked by Anthony Cuzzupe and Peter Wipf.

1. Procedure

Caution! Bromine is volatile and corrosive, and causes severe burns upon contact with skin. Proper protective equipment and an efficient fume hood are required.

A. ***D**-Ribonolactone.* A 1-L three-necked, round-bottomed flask fitted with a mechanical stirrer, a 100-mL pressure-equalizing addition funnel and an internal thermometer is charged with D-ribose (100 g, 0.67 mol), sodium bicarbonate (112 g, 1.3 mol, 2 equiv) (Note 1) and 600 mL of water. The mixture is stirred at room temperature for 15 min, during which time most of the solids dissolve (Note 2). The flask is then immersed in an ice-water bath. The addition funnel is filled with bromine (112 g, 0.70 mol, 1.04 equiv) (Note 1) and the bromine is added to the vigorously stirred aqueous solution at a rate of about 2 drops/sec such that the reaction temperature does not exceed 5°C. When the addition is complete (about 1 hr), the funnel is replaced with a stopper and the resulting orange solution is stirred for an additional 50 min. Sodium bisulfite (6.5 g, 62.5 mmol) is added in order to completely discharge the orange color (Note 3). The clear aqueous solution is transferred to a 2-L flask and evaporated on a rotary evaporator (bath temperature 60°C–70°C, water aspirator pressure) until a wet slurry remains.

Absolute EtOH (400 mL) and toluene (100 mL) are added to give a cloudy suspension and the solvent is removed by rotary evaporation (bath temperature 50°C, water aspirator pressure) to provide a damp solid. Absolute EtOH (400 mL) is added and the mixture is heated on a steam bath for 30 min. The hot ethanolic suspension is filtered and the solids are rinsed with hot absolute EtOH (100 mL). The filtrate is cooled to room temperature, and then refrigerated for 16 hr. The crystalline product is filtered, rinsed first with cold absolute EtOH (100 mL), then with Et_2O (100 mL), and dried under vacuum (room temperature, 0.25 mmHg) to yield 125 g of crude product (Note 4). The filtrate is concentrated (200 mL) and refrigerated to obtain additional product, which is filtered, washed with cold EtOH (25 mL) and Et_2O (25 mL) and dried under vacuum (room temperature, 0.25 mmHg) to provide 35 g of additional crude product.

B. *2,3-Isopropylidene-(**D**-ribonolactone).* In a 2-L round-bottomed flask, the crude ribonolactone (160 g) from above is suspended in 700 mL of dry acetone, 100 mL of 2,2-dimethoxypropane and conc. H_2SO_4 (1 mL, 20 mmol) (Note 1). The solution is stirred vigorously at room temperature for 50 min, then silver carbonate (20 g, 73 mmol) is added (Notes 1 and 5). The resulting suspension is stirred at room temperature for 50 min, then the suspension is filtered through a 2-cm Celite pad. The solids are rinsed with acetone (100 mL), and the filtrate is evaporated to dryness. The crude acetonide is dissolved in EtOAc (250 mL) with heating on a steam bath. The resulting suspension is filtered through a 2-cm Celite pad, the solids are rinsed with hot EtOAc (50 mL), and the filtrate is allowed to cool to room temperature. Crystals of 2,3-isopropylidene-(**D**-ribonolactone) form spontaneously as the solution cools. The resulting crystalline product is filtered and dried under vacuum (room temperature, 0.25 mmHg) to yield 68.7 g of product. The mother liquor is concentrated to 50 mL to yield 22.5 g of additional product after filtration and drying as above. The solids are combined to afford 91.2 g (73% overall yield from ribose) of white crystalline solid (Notes 6 and 7).

2. Notes

1. Ribose, bromine and silver carbonate were purchased from Acros Organics. Sodium bicarbonate was obtained from Mallinckrodt Baker, Inc. 2,2-Dimethoxypropane was obtained from Avocado Research Chemicals, Ltd. Deionized water was used as the solvent. All reagents were used as purchased.

2. Some of the sodium bicarbonate remains undissolved, but this does not affect the reaction yield, as the excess will be consumed during the course of the reaction.

3. If any orange color remains at this stage, the product will develop a pale brown color that cannot be removed by activated carbon.

4. The major contaminant is ~40-45% sodium bromide. The submitters recrystallized the crude product twice from hot *n*-BuOH (800 mL) to obtain 80 g of pure material. However, the crude material is of sufficient quality for the subsequent reaction. Pure ribonolactone has the following physical properties: mp 85-87°C; $[\alpha]_D^{25}$+11.9 (*c* 0.99, H_2O); IR (KBr) cm^{-1}: 3513, 3373, 3159, 1761, 1627, 1397, 1197, 1143; R_f 0.56 (30% v/v MeOH/$CHCl_3$, *p*-anisaldehyde in ethanol stain); ^{1}H NMR (300 MHz, DMSO-d_6) δ: 3.56 (dd, 2 H, J = 5.4, 3.6 Hz), 4.11 (app t, 1 H, J = 4.9 Hz), 4.21 (t, 1 H, J = 3.5 Hz), 4.40 (dd, 1 H, J = 7.7, 5.4 Hz), 5.15 (t, 1 H, D_2O exch, J = 5.4 Hz), 5.35 (d, 1 H, D_2O exch, J = 3.8 Hz), 5.73 (d, 1 H, D_2O exch, J = 7.7 Hz); ^{13}C NMR (75 MHz, DMSO-d_6) δ: 60.5, 68.6, 69.3, 85.4, 176.5.

5. As an alternative to the expensive silver carbonate, the submitters also used 30 g of the weakly basic resin Amberlyst A-21. In this instance, additional acetone (200 mL total) is used to rinse the filtered solids, and care must be taken upon evaporation of the solvent to remove the residual water.

6. The recrystallized **D**-ribonolactone acetonide has the following physical properties: mp 134-137°C; $[\alpha]_D^{24}$-66.7 (*c* 1.03, $CHCl_3$); IR (KBr) cm^{-1}: 3469, 2991, 2952, 2932, 1767, 1467, 1389, 1379, 1224; R_f 0.51 (1:3, hexane:EtOAc; *p*-anisaldehyde in ethanol stain); ^{1}H NMR (300 MHz, DMSO-d_6) δ: 1.30 (s, 3 H); 1.34 (s, 3 H), 3.67-3.55 (m, 2 H), 4.60 (t, 1 H, J = 2.1 Hz), 4.76 (s, 2 H), 5.30 (t, 1 H, D_2O exch, J = 5.0 Hz); ^{13}C NMR (75 MHz, DMSO-d_6) δ: 25.1, 26.6, 60.4, 75.0, 78.1, 82.3, 111.6, 174.3; Anal. Calcd for $C_8H_{12}O_5$: C, 51.06; H, 6.43. Found: C, 51.26; H, 6.46.

7. Extremely pure ribonolactone acetonide melts at >140°C, but material with a melting point >130°C is suitable for most reactions. Very pure material can be obtained by 1-2 additional recrystallizations from hot EtOAc.

Safety and Waste Disposal Information

All hazardous materials should be handled and disposed of in accordance with "Prudent Practices in the Laboratory"; National Academy Press; Washington, DC, 1995.

3. Discussion

The present procedure represents a modification of a previously published procedure for the bromine oxidation of ribose to ribonolactone.[3] The improved procedure allows for the addition of a liquid reagent (bromine) instead of the solid reagent (sodium carbonate) used in the previous preparation, and thus requires neither a powder addition apparatus nor constant attention. Additionally, the present procedure does not require the careful monitoring of pH, the absence of which can lead to the complete failure of the previous preparation.

Industrial preparations of ribonolactone involve the epimerization of arabinonic acid salts under very caustic and harsh conditions, followed by fractional crystallization of the resulting ribonic acid salts from the arabinonic/ribonic acid salt mixture, and the cyclization of the resulting salts.[4-6] An alternative rhenium-catalyzed oxidation procedure requires the preparation of an expensive catalyst and the removal of substantial amounts of benzalacetone (used as a terminal oxidant) and its reduction products.[7]

D-Ribonolactone is no longer commercially available in large quantities, and is very expensive. The acetonide is also very expensive. Although the above preparation of D-ribonolactone and D-ribonolactone acetonide is not high-yielding, the starting materials are inexpensive and the preparation is quite convenient. D-Ribonolactone and its derivatives have been used for the syntheses of many nucleoside analogs[8,9] and natural products.[10,11]

1. Department of Medicinal Chemistry, University of Michigan, 428 Church St., Ann Arbor, MI 48105.
2. BioCryst Pharmaceuticals, 2190 Parkway Lake Dr., Birmingham, AL 35244.
3. Pudlo, J. S.; Townsend, L. B. In *Nucleic Acid Chemistry: Improved and New Synthetic Procedures, Methods, and Techniques, Part 4*.; Townsend, L. B., Tipson, R. S., Eds.; John Wiley & Sons: New York, **1991**, p. 51-53.
4. Flexser, L. A.; Hoffmann-La Roche, Inc.: US Patent 2,438,883, **1948**.
5. Sternbach, L. H.; Hoffmann-La Roche, Inc.: US Patent 2,438,881, **1948**.
6. Schmidt, W.; Paust, J.; BASF Aktiengesellschaft: US Patent 4,294,766, **1981**.

7. Isaac, I.; Stasik, I.; Beaupere, D.; Uzan, R. *Tetrahedron Lett.* **1995**, *36*, 383-386.
8. Pankiewicz, K. W.; Sochacka, E.; Kabat, M. M.; Ciszewski, L. A.; Watanabe, K. A. *J. Org. Chem.* **1988**, *53*, 3473-3479.
9. Cheng, J. C.-Y.; Hacksell, U.; Daves, G. D. J. *J. Org. Chem.* **1985**, *50*, 2778-2780.
10. Fürstner, A.; Radkowski, K.; Wirtz, C.; Goddard, R.; Lehmann, C. W.; Mynott, R. *J. Am. Chem. Soc.* **2002**, *124*, 7061-7069.
11. Jiang, S.; Mekki, B.; Singh, G.; Wightman, R. H. *Tetrahedron Lett.* **1994**, *35*, 5505-5508.

Appendix
Chemical Abstract Nomenclature (Registry Number)

D-Ribose; (50-69-1)

Sodium bicarbonate: Carbonic acid monosodium salt; (144-55-8)

Bromine; (7726-95-6)

Sodium bisulfite: Sulfurous acid, monosodium salt; (7631-90-5)

D-Ribonolactone: D-Ribonic acid, γ-lactone; (5336-08-3)

2,2-Dimethoxypropane: Propane, 2,2-dimethoxy-; (77-76-9)

Silver carbonate: Carbonic acid, disilver(1+) salt; (534-16-7)

PREPARATION OF 4-ACETYLAMINO-2,2,6,6-TETRAMETHYLPIPERIDINE-1-OXOAMMONIUM TETRAFLUOROBORATE, AND THE OXIDATION OF GERANIOL TO GERANIAL

(2,6-Octadienal, 3,7-dimethyl-, (2*E*)-)

A. NHAc; 2.1 HBF_4; NHAc, BF_4^- + NHAc, H OH, BF_4^-; 1.0 NaOCl; 2 NHAc, BF_4^- + NaCl + H_2O

B. OH + NHAc, BF_4^-; CH_2Cl_2, silica gel; O, H + NHAc, H OH

Submitted by James M. Bobbitt and Nabyl Merbouh.[1]
Checked by Peter Wipf and David Amantini.

1. Procedures (Note 1)

A. *4-Acetylamino-2,2,6,6-tetramethylpiperidine-1-oxoammonium tetrafluoroborate*. In a 250-mL one-necked, round-bottomed flask equipped with a 3-cm magnetic stirring bar, 4-acetylamino-2,2,6,6-tetramethyl-1-piperidinyloxy (4-acetamido-TEMPO) (25.0 g, 0.117 mol) (Note 2) and 50 mL of water are added. An aqueous solution of HBF_4 (48% aqueous solution, 17.7 mL, 0.135 mol) (Note 3) is charged into a 25-mL dropping funnel and added dropwise over 30 min to the vigorous stirring orange mixture. A dark brown solution is formed initially, followed by formation of a yellow precipitate. The yellow slurry is stirred for an additional 30 min. Commercial sodium hypochlorite (Chlorox® bleach, 6.00 % NaOCl) (65.5 mL, 0.058 mol) is transferred into a 100-mL dropping funnel and added dropwise over a 1 hr period to the heterogeneous mixture (Note 4). The yellow slurry is then cooled to 0°C and stirred at this temperature for 2 hrs. The mixture is filtered and the yellow solid washed with cooled water (4°C, 2 x 20 mL) (Note 5) and dichloromethane (3 x 20 mL) to remove sodium chloride and unreacted 4-acetamido-TEMPO, respectively. The bright yellow salt is compressed with a spatula to remove additional solvent, transferred into a 100-mL round-bottomed flask and dried under high vacuum at room temperature overnight (Note 6). The product is obtained as a bright yellow solid (27.4 g, 78%, 98% purity) (Note 7).

B. The oxidations of a number of alcohols with 4-acetylamino-2,2,6,6-tetramethylpiperidine-1-oxoammonium perchlorate have been described on a 10-mmol scale.[2] The oxoammonium perchlorate salt detonated when a 9 g sample was dried at 40°C under high vacuum, after being apparently stable for several years.[3] Thus, attention has been shifted to development of the oxoammonium tetrafluoroborate salt. In all cases investigated, the oxidative properties of the two salts are identical.

Oxidation of geraniol to geranial. Into a 500-mL one-necked, round-bottomed flask are added geraniol (7.70 g, 50.0 mmol) (Note 8) and 400 mL of dichloromethane. 4-Acetylamino-2,2,6,6-tetramethylpiperidine-1-oxoammonium tetrafluoroborate (16.5 g, 55.0 mmole, 1.10 eq) and 2 g of silica gel are added to the solution (Note 9). The flask is equipped with a mechanical stirrer and the reaction mixture is vigorously stirred at room temperature for 6 hr. The reaction begins immediately, and the yellow slurry turns gradually to the off-white color of the reduced oxidant. A 1-cm thick pad of silica gel is placed in a fritted glass funnel (*ca.* 7 cm in diameter), wetted with dichloromethane and covered with a piece of filter paper. The slurry is carefully poured onto the silica gel pad and filtered (Note 10). The pad is washed with four successive 50 mL portions of dichloromethane (Note 11). The solvent is evaporated under vacuum (room temperature) to give 7.08 g (93%) of geranial (98% purity by GC analysis) as a colorless oil (Note 12). The hydroxylamine-silica gel precipitate can be processed to obtain 4-acetamido-TEMPO for recycling (Note 13).

2. Notes

1. No professional safety check has been carried out to determine the stability of 4-acetylamino-2,2,6,6-tetramethylpiperidine-1-oxoammonium

tetrafluoroborate salt. All the procedures herein described have been carried out behind a safety shield.

2. 4-Acetamido-TEMPO was purchased from TCI and used without further purification; however, submitters report that it can readily be prepared from 4-amino-2,2,6,6-tetramethylpiperidine (Fluka). In their original procedure for the preparation of 4-acetamido-TEMPO,[2] potassium carbonate was used to basify 4-acetylamino-2,2,6,6-tetramethylpiperidinium acetate; however, due to its insolubility the potassium tetrafluoroborate can accumulate in the salt. Therefore, sodium carbonate should be used and all potassium salts should be avoided. Sodium tetrafluoroborate is quite water-soluble.

3. Tetrafluoroboric acid (48% aqueous solution) was purchased from ACROS.

4. Slow addition is necessary since the tetrafluoroborate oxoammonium salt reacts with excess bleach to produce undesired byproducts.

5. The oxoammonium tetrafluoroborate salt is fairly soluble in pure water (6 g/100 mL at 0°C, 8 g/100 mL at 20°C and >100 g/100 mL at 100°C), so care must be taken during the washes.

6. The oxoammonium tetrafluoroborate salt reacts, albeit slowly, with hot water. If the wet salt is dried in an oven, some decomposition will take place.

7. The melting point of the oxoammonium tetrafluoroborate salt is a vigorous, instantaneous decomposition in a capillary tube: mp 184-185°C (dec.). The submitters report that the observed melting point can vary from 180°C to 194°C, even though the purity is about the same. Since the melting points are ambiguous, the purity of the salt was measured by oxidation of a known quantity of 1-decanol with a known, limited amount of salt. The conversion of 1-decanol to 1-decanal was quantified using ^{1}H NMR (300 MHz, $CDCl_3$). The C-2 proton NMR signal (triplet) at 2.43 ppm (decanal) and the C-1 proton signal (triplet) at 3.65 ppm (1-decanol) are integrated and the percentage of conversion is compared with the theoretical value. In a 25-mL round-bottomed flask 1-decanol (0.079 g, 0.50 mmol), dichloromethane (7.0 mL), silica gel (0.10 g) and 4-acetylamino-2,2,6,6-tetramethylpiperidine-1-oxoammonium tetrafluoroborate (0.075 g, 0.25 mmol, 0.50 eq) were added and the resulting slurry left under stirring at room temperature for 12 hr. The mixture was filtered through a 1-cm silica gel pad and subsequently washed with dichloromethane (25 mL). The solvent was removed under vacuum at room temperature to give a mixture of 1-decanol and decanal as a colorless oil.

8. Geraniol (99% purity) was purchased from ACROS and used without further purification.

9. The amount of silica gel and the reaction time depend on the alcohol being oxidized. Geraniol, an allylic alcohol, requires less silica gel and takes less time. Primary aliphatic alcohols will require twice as much silica gel and longer reaction times.[2]

10. If the slurry is filtered without the pad of clean silica gel, a small amount of nitroxide will contaminate the product; however, some product loss occurs due to its incomplete removal from the silica gel.

11. The submitters report that this solution can be used for various reactions, without actual isolation of the carbonyl compound. Reactions reported include Wittig, Grignard, and Baylis-Hillman reactions, as well as cyanohydrin and acetal formation.

12. The crude aldehyde can be distilled at 20 mm and 115-120°C (lit.[4] 117°C at 20 mm) to give 6.53 g (89%) of geranial with spectral and chromatographic properties identical with the undistilled product: colorless oil, IR (KBr) [cm^{-1}]: 2967, 2918, 2836, 1676, 1633, 1444, 1381, 1194, 1121; EI-MS: *m/z* (%) = 153 (7), 152 (24), 137 (19), 123 (19), 109 (29), 94 (39), 84 (51), 69 (100); ^{1}H NMR (300 MHz, $CDCl_3$, TMS): δ 1.62, (s, 3 H), 1.69 (s, 3 H), 2.18 (s, 3 H), 2.21-2.28 (m, 4 H), 5.05-5.10 (m, 1 H), 5.87-5.91 (m, 1 H), 10.00 (d, J = 8.0 Hz, 1 H); ^{13}C NMR (75 MHz, $CDCl_3$) δ: 17.4, 17.6, 25.5, 25.6, 40.5, 122.5, 127.3, 132.7, 163.5, 191.0. GC-Analysis was performed with a EC-1 column of 30 m /0.32 mm ID and a T-ramp of 70°C to 240°C in 15°C/min, providing a retention time for geraniol of 6.94 min.

13. The submitters report that the spent oxidant consisting of a mixture of hydroxylamine tetrafluoroborate and silica gel can be collected from several oxidations and processed as follows to give 4-acetamido-TEMPO for recycling: The mixture is washed with several portions of warm water to extract hydroxylamine tetrafluoroborate salt from the silica gel. The aqueous solution is basified with $NaHCO_3$ to pH 7 and treated with an excess of H_2O_2 (30% aqueous solution). The 4-acetamido-TEMPO can be extracted with dichloromethane. Recrystallization from boiling water or boiling ethyl acetate provides purified 4-acetamido-TEMPO, which tends to be quite soluble at the boiling point of either solvent and precipitates quickly upon cooling. The resulting 4-acetamido-TEMPO should melt above 144°C to be judged suitable for further use.

Safety and Waste Disposal Information

All hazardous materials should be handled and disposed of in accordance with "Prudent Practices in the Laboratory"; National Academy Press; Washington, DC, 1995. If all of the materials are recycled as described, the only waste products are NaCl and $NaBF_4$.

3. Discussion

Oxoammonium salts are stable, non heavy-metal, specific oxidizing agents for the preparation of aldehydes or ketones from alcohols.[5] Under normal conditions,[2] the reactions are nearly quantitative; the reactions are colorimetric; and product isolation is simple. Anhydrous conditions may be used, but are not necessary.

Oxoammonium salt oxidations are actually the stoichiometric version of nitroxide-catalyzed oxidations using a secondary oxidant such as bleach, as described in a previous *Organic Syntheses* procedure.[6] Disadvantages of the catalyzed oxidations are that a two phase system is generally used and that the secondary oxidant can cause undesired reactions. Catalytic reactions are, however, more suitable for large-scale reactions. On the other hand, stoichiometric oxidations undergo fewer side reactions, but are more appropriate on a 1-50 mmol scale.

Other than convenience, two major advantages of oxoammonium salt reactions are that phenolic alcohols can be oxidized without protecting the phenol[2,7] and allylic alcohols can be oxidized without isomerization of double bonds.[2, 8, 9]

Many oxoammonium salts are known with various substitution patterns and different anions.[5] The properties vary tremendously, being especially important with respect to their hygroscopic nature and their rates of reaction. 4-Acetylamino-2,2,6,6-tetramethylpiperidine-1-oxoammonium perchlorate and tetrafluoroborate salts were described in our preliminary publication,[2] and the perchlorate was used exclusively for the oxidations. However, the perchlorate proved to be unstable and its use was discontinued.[3] All further work has been done with the tetrafluoroborate oxoammonium salt, which is a bright yellow crystalline material, completely non-hygroscopic and stable indefinitely. Its solubility and oxidation properties appear to be identical with those of the perchlorate.

The tetrafluoroborate salt possesses sufficient solubility in dichloromethane to react easily, and its reduced form, 4-acetylamino-2,2,6,6-*N*-hydroxypiperidinium tetrafluoroborate is colorless and essentially insoluble in this solvent. Thus, it can be removed by filtration, leaving a quantitative yield of aldehyde or ketone in solution.

Reaction rates vary, allylic and benzylic alcohols being fast (1-3 hr), acetylenic and secondary alcohols next (4-6 hr) and primary aliphatic alcohols being slowest (2-3 days without silica gel).[2] Silica gel is an effective catalyst and may also assist in the purification of the products. When silica gel is used, primary aliphatic alcohols are oxidized in 10-15 hr and allylic and benzylic alcohols require a lower catalyst loading. Completion of the oxidation can be easily monitored colorimetrically, since the yellow tetrafluoroborate oxoammonium salt starting material is converted to its white reduced form.

The most serious side reaction is the rapid oxidation of amines, a reaction that is not well understood.[2,5] Two slow competing reactions, the oxidation of benzyloxy groups[10] and the addition to activated double bonds,[2,11] can result in side product formation, particularly when slowly oxidized substrates (primary aliphatic alcohols) are used. Several other side reactions have been documented for oxoammonium salts,[5] but most are slow and of lesser importance, especially when the anion is tetrafluoroborate.

For reasons that are not understood, alcohols with a β-oxygen (in any functional group) or β-nitrogen (as amide) react so slowly with the oxidant as to be useless. Interestingly, this limitation is apparently not true for nitroxide-catalyzed reactions.[6]

Oxidations in dichloromethane are slightly acidic, due to the reduced oxidant, which has a pKa of about 5.6.[2] For this reason, simple acetal groups and such acid labile protecting groups as the *tert*-butyldimethylsiloxy group are slowly cleaved, although the *tert*-butyldiphenylsiloxy group seems to be stable.

Stoichiometric oxidations can be carried out in the presence of a base such as pyridine, although only one paper has appeared on the subject.[12] The reaction takes place in a different manner with two equivalents of oxidant and pyridine being required. The products are the desired oxidation product, pyridine tetrafluoroborate, and nitroxide. Under these conditions, acetal groups are stable, and b-oxygenated materials are oxidized.

1. Department of Chemistry, University of Connecticut, Storrs, CT 06269-3060.
2. Bobbitt, J. M. *J. Org. Chem.* **1998**, *63*, 9367.
3. Bobbitt, J. M. *Chem. & Eng. News*, July 19, **1999**, *77*, 6.
4. Cardillo, G.; Orena, M.; Sandri, S. *Synthesis*, **1976**, 394.

5. For reviews, see (a) Bobbitt, J. M.; Flores, M. C. L. *Heterocycles* **1988**, *27*, 509. (b) de Nooy, A. E. J.; Besemer, A. C.; van Bekkum, H. *Synthesis* **1996**, 1153. (c) Merbouh, N.; Bobbitt, J. M.; and Brückner, C. *Org. Prep. Proced. Int.* **2004**, *36,* 1.
6. Anelli, P. L.; Montanari, F.; Quici, S. *Org. Synth.* **1990**, *69*, 212.
7. Abad, A.; Agullo, C.; Cunat, A. C.; Perni, R. H. *Tetrahedron: Asymmetry*, **2000**, *11*, 1607.
8. Koch, T.; Hoskovec, M.; Boland, W. *Tetrahedron* **2002**, *58*, 3271.
9. Hoye, T. R.; Hu, M. *J. Am. Chem. Soc.* **2003**, *125*, 9576.
10. Miyazawa, T.; Endo T. *Tetrahedron Lett.* **1986**, *27*, 3395.
11. Takata, T.; Tsujino, Y.; Nakanishi, S.; Nakamura, K.; Yoshida, E.; Endo, T. *Chem. Lett.* **1999**, 937.
12. Merbouh, N.; Bobbitt, J. M.; Brückner C. *Tetrahedron Lett.* **2001**, *42*, 8793.

Appendix
Chemical Abstract Nomenclature (Registry Number)

4-Acetylamino-2,2,6,6-tetramethylpiperidine-1-oxoammonium tetrafluoroborate: Piperidinium, 4-(acetylamino)-2,2,6,6-tetramethyl-1-oxo-, tetrafluoroborate(1-) (9); (219543-09-6)

4-Acetylamino-2,2,6,6-tetramethylpiperidine-1-oxoammonium perchlorate: Piperidinium, 4-(acetylamino)-2,2,6,6-tetramethyl-1-oxo-, perchlorate (9); (219543-08-5)

4-Acetamido-TEMPO: 1-Piperidinyloxy, 4-(acetylamino)-2,2,6,6-tetramethyl- (9); (14691-89-5)

Tetrafluoroboric acid: Borate(1-), tetrafluoro-, hydrogen (8,9); (16872-11-0)

Sodium hypochorite (Clorox®): Hypochlorous acid, sodium salt (8,9); (7681-52-9)

1-Decanol: 1-Decanol (9); (112-30-1)

Decanal: Decanal (8,9); (112-31-2)

Hydrogen peroxide: Hydrogen peroxide (9); (7722-84-1)

4-Amino-2,2,6,6-tetramethylpiperidine: 4-Piperidinamine, 2,2,6,6-tetramethyl- (9); (36768-62-4)

4-Acetylamino-2,2,6,6-tetramethylpiperidinium acetate: Acetamide, *N*-(2,2,6,6-tetramethyl-4-piperidinyl)-, monoacetate (9); (136708-43-5)

Geraniol: 2,6-Octadien-1-ol, 3,7-dimethyl-, (2*E*)-; (106-24-1)

Geranial: 2,6-Octadienal, 3,7-dimethyl-, (2*E*)-; (141-27-5)

(2*S*)-(–)-3-*exo*-(MORPHOLINO)ISOBORNEOL [(–)-MIB] ([1*R*-(*exo*,*exo*)]-1,7,7-Trimethyl-3-morpholin-4-yl-bicyclo[2.2.1] heptan-2-ol)

A. 1. Potassium *t*-butoxide; 2. *i*-Amylnitrite

B. $LiAlH_4$

C. $(BrCH_2CH_2)_2O$, Et_3N, DMSO

Submitted by Young K. Chen,[1] Sang-Jin Jeon,[1] Patrick J. Walsh,[1] and William A. Nugent.[2]
Checked by Christopher Kendall and Peter Wipf.

1. Procedure

A. *Camphorquinone oxime*. A flame-dried 1-L three-necked, round-bottomed flask equipped with a mechanical stirrer, a thermometer and septum is charged with potassium *t*-butoxide (46.0 g, 410 mmol) (Note 1). The contents of the flask are thoroughly purged with a stream of N_2 exhausted through an oil bubbler. After 15 min, 500 mL of Et_2O (Note 2) is added *via* cannula and the flask is submerged in a cold bath cooled to -30°C (Note 3). A second 250-mL round-bottom flask is charged with (*R*)-camphor (50.0 g, 328 mmol) (Note 4) and 100 mL of Et_2O. The resulting clear solution is added *via* cannula into the first flask over 10 min while keeping the internal temperature below -30°C. The second flask is thoroughly rinsed with 20 mL of Et_2O, which is also transferred into the first flask. The cooling bath is removed and the reaction mixture is allowed to warm to room temperature. After stirring at room temperature for 30 min, the reaction mixture is cooled to -30°C. Isoamyl nitrite (55.0 mL, 409 mmol) (Note 5) is added by syringe over 20 min while keeping the internal temperature below -30°C. An orange to red color appears during the addition of isoamyl nitrite. The reaction mixture is allowed to warm to room temperature and is stirred for 16 hours at ambient temperature under a N_2 atmosphere. The solution is extracted with water (3 x 150 mL), and the combined aqueous layers (approximately 450 mL) are cooled in an ice bath with magnetic stirring. The pH is adjusted to 4 by dropwise addition of approximately 30 mL of conc. HCl. After the pH is adjusted, an off-white solid precipitates from solution. The biphasic mixture is extracted with CH_2Cl_2 (3 x 150 mL). The combined organic layers are washed successively with 50 mL of saturated $NaHCO_3$, 50 mL of water, 50 mL of brine, and then dried over anhydrous $MgSO_4$. The solvent is removed *in vacuo* to afford 52.8 g (89%) of the title compound. The solid is a 10/90 mixture of *syn-* and *anti-*camphorquinone oximes (Note 6). This material is suitable for the next reaction without further purification.

B. *(2S)-(-)-3-exo-Aminoisoborneol.* A 250-mL two-necked, round-bottomed flask is charged with camphorquinone oximes (12.1 g, 66.8 mmol). The flask is fitted with a condenser and a septum, and is thoroughly purged with a steady stream of N_2. After 15 min, the N_2 flow is reduced to a slow bleed and 30 mL of anhydrous THF (Note 2) is added via syringe. The homogeneous solution is cooled in an ice bath to 0°C. A 1.0 M solution of $LiAlH_4$ in THF (100 mL, 100 mmol) (Note 7) is slowly transferred *via* cannula to the mixture over 30 min (Note 8). After vigorous H_2 gas evolution ceased, the reaction mixture is allowed to warm to ambient temperature and then heated at reflux for 30 min. The solution is cooled to room temperature, diluted with 65 mL of Et_2O, cooled to 0°C and quenched by the successive dropwise addition of 3.8 mL of water, 3.8 mL of 10% NaOH solution, and 11.4 mL of water (Note 9). The colorless precipitate was vacuum-filtered through Celite, and the filter cake was washed with THF (3 x 20 mL) (Note 10). The combined filtrate was concentrated to give 10.4 g (92%) of a waxy solid. This material was used in the next step without further purification.

C. *(2S)-(-)-exo-(Morpholino)isoborneol.* To a 150-mL round-bottomed flask charged with (*2S*)-(-)-3-*exo*-aminoisoborneol (6.53 g, 38.6 mmol) is added 40 mL of reagent grade DMSO followed by Et_3N (16.3 mL, 117 mmol) (Note 11). Bis(2-bromoethyl) ether (6.50 mL, 46.5 mmol) (Note 11) in 30 mL of DMSO is added drop-wise over 10 min and the reaction mixture is stirred at ambient temperature under N_2 for 72 hours. The solution is poured into 400 mL of water, and the aqueous mixture is extracted with Et_2O (3 x 150 mL). The combined organic layers are washed successively with 100 mL of water, 50 mL of brine and then dried over anhydrous

$MgSO_4$. The solvent is removed *in vacuo* and the residue is purified by flash column chromatography (Notes 12, 13 and 14) to give 5.23 g (57%) of (-)-MIB as a colorless solid.

2. Notes

1. Potassium *t*-butoxide was purchased from Acros and used under a stream of N_2.

2. Et_2O and THF were purchased from Fisher and dried by passage through an activated alumina column under N_2.

3. The reaction was successfully performed at temperatures ranging from –30°C to –50°C.

4. (1*R*)-(+)-Camphor 98%, was purchased from Aldrich Chemical. The optical purity of (1*R*)-(+)-camphor varies with the natural source. For analysis of the optical purity of commercial (+)- and (–)-camphor see: Armstrong, D. W.; Lee, J. T.; Chang, L. W. *Tetrahedron: Asymmetry* **1998**, *9*, 2043.

5. Isoamyl nitrite was purchased form Aldrich Chemical Company, Inc. Because of the appalling odor of the reagent, it was used without prior distillation.

6. *Anti*-(1*S*)-(–)-camphorquinone 3-oxime ^1H NMR (500 MHz, $CDCl_3$) δ: 0.89 (s, 3 H), 1.03 (s, 3 H), 1.00 (s, 3 H), 1.62-1.52 (m, 2 H), 1.83-1.74 (m, 1 H), 2.08-2.00 (m, 1 H), 3.25 (d, $J = 4.3$, 1 H), 9.70-8.80 (br, 1 H, N-OH); ^{13}C {^1H} NMR (125 MHz, $CDCl_3$) δ: 9.3, 18.0, 21.1, 24.2, 31.1, 45.3, 47.1, 58.8, 160.1, 204.5. *Syn*-(1*S*)-(–)-camphorquinone 3-oxime ^1H NMR (500 MHz, $CDCl_3$) δ: 0.93 (s, 3 H), 1.01 (s, 3 H), 1.03 (s, 3 H), 1.66-1.55 (m, 2 H), 1.83-1.74 (m, 1 H), 2.14-2.08 (m, 1 H), 2.70 (d, $J = 4.2$, 1 H), 9.70-8.80 (br, 1 H, N-OH); ^{13}C {^1H} NMR (125 MHz, $CDCl_3$) δ: 8.8, 18.4, 21.0, 25.4, 30.4, 47.4, 50.0, 60.0, 156.6, 205.3.[3,4a]

7. 1.0 M $LiAlH_4$ in THF was purchased from Aldrich. The yields and diastereoselectivity were much higher when homogeneous solutions of $LiAlH_4$ were used instead of the powder form.

8. H_2 gas evolution was vigorous at the beginning of the addition.

9. Addition of the initial 3.8 mL of water is accompanied by vigorous generation of H_2 and is very exothermic.

10. The best product recovery was accomplished when the filter cake was washed with THF. For spectral properties see reference 4b.

11. Di(bromoethyl) ether and Et_3N were purchased from Aldrich. Reagent grade DMSO was purchased from Fisher.

12. ~200 g of silica, 10% to 15% EtOAc in hexanes. TLC in 20% EtOAc in hexanes; R_f 0.25; Stained brown in I_2 Chamber.

13. Recrystallization is also possible from hexanes 4 mL/g at –30°C.[5]

14. Characterization data for (2*S*)-(–)-*exo*-(morpholino)isoborneol: $[\alpha]^{20}_D$ = -6.9 (*c* = 1.0, MeOH); mp = 65-66°C (hexane); ^{1}H NMR (400 MHz, C_6D_6) δ: 0.68 (s, 3 H); 0.77-0.70 (m, 1 H), 0.92-0.82 (m, 1 H), 1.02 (s, 3 H), 1.13 (s, 3 H), 1.35 (td, 1 H, *J* = 12.2, 3.6), 1.51 (tt, 1 H, *J* = 12.0, 4.6 Hz), 1.66 (d, 1 H, *J* = 4.7 Hz), 2.07 (d, 1 H, *J* = 7.1 Hz), 2.10 (bs, 2 H), 2.30 (bs, 2 H), 3.40 (bs, 4 H), 3.43 (d, 1 H, *J* = 7.1 Hz), 3.91 (s, 1 H). ^{13}C {^{1}H} NMR (76 MHz, C_6D_6, d_l = 5 sec) δ: 12.4, 21.7, 22.8, 28.6, 33.2, 46.0, 47.2, 50.1, 67.4, 74.0, 79.6. IR (KBr) 3460, 3367, 1478, 1448, 1396, 1360, 1284, 1261, 1202, and 1200 cm^{-1}; EIMS *m/z* 239 (M^+, 13), 154 (100); HRMS (EI) *m/z* calcd for $C_{14}H_{25}NO_3$: 239.1885, found 239.1889. Enantiomeric excess was determined as followed: To a screw cap vial (1 dram) charged with (–)-MIB (24 mg, 0.1 mmol) was added dichloromethane (1 mL), followed by Et_3N (17 μL, 0.12 mmol), and DMAP (~2 mg). *p*-Bromobenzoyl chloride (22 mg, 0.1 mmol) was added to the clear solution and stirred for 10 min. The reaction mixture was concentrated under reduced pressure and the residue purified by column chromatography (5% ethyl acetate in hexanes). HPLC analysis of the resultant *p*-bromobenzyl ester established the enantiomeric excess as 96.0% ee (Chiralcel OD column, flow 1 mL/min, 254 nm, 2% isopropanol in hexanes; minor isomer 4.6 min, and major isomer 5.3 min).

Safety and Waste Disposal Information

All hazardous materials should be handled and disposed of in accordance with "Prudent Practices in the Laboratory"; National Academy Press; Washington, DC, 1995.

3. Discussion

(–)-MIB has been shown to be an excellent chiral ligand in the asymmetric alkylation of aldehydes. In comparison, MIB was equal or better than DAIB in Et_2Zn addition to aromatic aldehydes with ee's > 95. Highly enantioselective Et_2Zn additions to alkyl substituted aldehydes are also possible with (–)-MIB.[5] Remarkably, the generation of (*S*)-1-phenylpropan-1-ol with 90% ee can be achieved when MIB of only 10% ee was used in the Et_2Zn addition with benzaldehyde. This result shows a large positive non-linear effect that parallels DAIB.[6] Recently, MIB has been shown to be equally effective with DAIB in the alkenylzinc addition to aldehydes pioneered by Oppolzer.[7] A number of terminal alkynes with various substituents were used in the study. Using MIB in this reaction constitutes a powerful and practical method to access both enantiomers of allylic alcohols in high optical purity.[8] Furthermore, it has recently been shown that MIB can be used in a one-pot tandem asymmetric addition/diastereoselective epoxidation sequence to generate epoxy alcohols with up to three stereocenters with high enantio- and diastereoselectivity.[9]

The distinct advantage of MIB is its ease of preparation. Gram quantities of both enantiomers of MIB can be made in only three steps and one purification, while the most efficient synthesis of (–)-DAIB was achieved in six steps and involved a low yielding and laborious purification step to remove the undesired diastereomer.[4] Furthermore, MIB is a crystalline solid and can be stored for months in the presence of air without noticeable decomposition.[5]

1. Department of Chemistry, University of Pennsylvania, Philadelphia, PA, 19104-6323, USA.
2. Bristol-Myers Squibb Company, Process Research and Development Department, P. O. Box 4000, Princeton, New Jersey 08543-4000.
3. Roy, S.; Chakraborti, A. K. *Tetrahedron Lett.* **1998**, 6355-6356.
4. (a) White, J. D.; Wardrop, D. J.; Sundermann, K. F. *Org. Synth.* **2002**, 79, 125-129. (b) White, J. D.; Wardrop, D. J.; Sundermann, K. F. *Org. Synth.* **2002**, *79*, 130-138.
5. Nugent, W. A. *J. Chem. Soc., Chem. Comm.* **1999**, 1369-1370.
6. Chen, Y. K.; Costa, A. M.; Walsh, P. J. *J. Am. Chem. Soc.* **2001**, *123*, 5378-5379.
7. Oppolzer, W.; Radinov, R. N. *Helv. Chim. Acta* **1992**, *75*, 170-173.
8. (a) Chen, Y. K.; Lurain, A. E.; Walsh, P. J. *J. Am. Chem. Soc.* **2002**, *124*, 12225-12231. (b) Lurain, A. E.; Walsh, P. J. *J. Am. Chem. Soc.* **2003**, *125*, 10677-10683.
9. (a) Lurain, A. E.; Maestri, A.; Kelly, A. R.; Carroll, P. J.; Walsh, P. J. *J. Am. Chem. Soc.* **2004**, *126*, 13608-13609. (b) Lurain, A. L.; Carroll, P. J.; Walsh, P. J. *J. Org. Chem.* **2005**, *70*, 1262-1268.

Appendix
Chemical Abstracts Nomenclature (Registry Number)

anti-(1*R*)-(–)-Camphorquinone 3-oxime: 1,7,7-Trimethyl-bicyclo[2.2.1]heptane-2,3-dione 3-oxime; (31571-14-9)

(1*R*)-(+)-Camphor: 1,7,7-Trimethyl-bicyclo[2.2.1]heptan-2-one; (464-49-3)

(2*S*)-(–)-3-*exo*-Aminoisoborneol: 3-Amino-1,7,7-trimethyl-bicyclo[2.2.1]heptan-2-ol; (417199-73-7)

Bis(2-bromoethyl) ether: Ethane, 1,1'-oxybis[2-bromo-; (5414-19-7)

Lithium aluminum hydride; (16863-85-3)

Isoamyl nitrite: Isopentyl nitrite; (110-46-3)

Potassium *tert*-butoxide; (865-47-4)

Triethylamine: Ethanamine, *N,N*-diethyl-; (121-44-8)

CONVERSION OF ARYLALKYLKETONES INTO DICHLOROALKENES: 1-CHLORO-4-(2,2-DICHLORO-1-METHYLVINYL)BENZENE

(1-Chloro-4-(2,2-dichloro-1-methylethenyl)benzene)

A. 1-(4-chlorophenyl)ethanone + N_2H_4 → (ethanol, reflux) → 1-(4-chlorophenyl)ethanone hydrazone

B. 1-(4-chlorophenyl)ethanone hydrazone + CCl_4 → (CuCl, DMSO, aq NH_3) → 1-chloro-4-(2,2-dichloro-1-methylvinyl)benzene

Submitted by Valentine G. Nenajdenko,[1] Vasily N. Korotchenko,[1] Alexey V. Shastin,[2] and Elisabeth S. Balenkova.[1]
Checked by Kristin Brinner and Jonathan A. Ellman.

1. Procedure

A. *1-(4-Chlorophenyl)ethanone hydrazone.* A single-necked, 250-mL round-bottomed flask equipped with a reflux condenser is charged with 100 mL of absolute ethanol (Note 1), 100% hydrazine hydrate (9.7 mL, 0.2 mol) (Note 2) and 1-(4-chlorophenyl)ethanone (13.0 mL, 0.1 mol) (Note 3). The reaction solution is refluxed for 3 hr. The ethanol and excess of hydrazine are removed by evaporation at 50°C at 20 mm (Note 4). The slight-yellow crude 1-(4-chlorophenyl)ethanone hydrazone is redissolved in 10 mL of hot ethanol (Note 1). The mixture is cooled to –18°C, kept at that temperature overnight and filtered. The solid is washed with hexane (2 x 20 mL) (Note 5) and dried under vacuum (2-3 mm) for 2 hr to afford 13.6-14.8 g (81-88%) of colorless crystals (Note 6). The product is stored at –18°C until used in step B (Note 7).

B. *1-Chloro-4-(2,2-dichloro-1-methylvinyl)benzene.* A single-necked, 250-mL, round-bottomed flask equipped with a magnetic stirring bar and a pressure-equalizing dropping funnel is charged with 50 mL of dimethyl sulfoxide (DMSO) (Note 8), 1-(4-chlorophenyl)ethanone hydrazone (8.4 g, 0.05 mol) (Note 6) and 28% aqueous ammonium hydroxide (16.8 mL, 0.25 mol) (Note 9). Freshly purified copper (I) chloride (500 mg, 0.005 mol) (Note 10) was added in one portion. After 5 min, carbon tetrachloride (24 mL, 0.25 mol) (Note 11) is added dropwise over 30 min (Note 12), while the temperature is maintained between 15-20°C (water bath). After the addition is complete, the reaction mixture is stirred at room temperature overnight and quenched with 500 mL of 5% aqueous hydrochloric acid. The reaction products are extracted with hexane (4 x 50 mL) (Note 5). The combined extracts are dried over anhydrous sodium sulfate. After filtration and removal of the solvent by rotary evaporation (Note 13), the yellow crude product is purified by flash chromatography (5.0 x 5.0 cm) through silica gel (30 g, 63-200 mesh) using hexane (4-5 50 mL portions) (Notes 13 and 14) to afford 8.64-8.85 g (78-80% yield) of 1-chloro-4-(2,2-dichloro-1-methylvinyl)benzene as a colorless, viscous oil that is > 98% pure by ^{1}H NMR analysis (Note 15).

2. Notes

1. Reagent grade ethanol was used without further purification by the checkers.

2. 100% Hydrazine hydrate was purchased from E. Merck.

3. 1-(4-Chlorophenyl)ethanone (98% purity) was purchased from E. Merck. The checkers purchased 97% purity material from Aldrich.

4. The checkers report that ethanol was removed by rotary evaporation at 30 mm with the bath temperature at 30°C. Hydrazine was then removed at 10 mm with the bath temperature at 25°C. The submitters report that the ethanol–hydrazine mixture collected in a trapping flask can be used for the preparation of next portion of hydrazone.

5. Reagent grade hexane was used without further purification.

6. mp 48.4-50.1°C (the same melting point was observed after repeated recrystallizations; however, the submitters report a mp of 55-56 °C). ^{1}H NMR (400 MHz, DMSO-d_6) δ: 2.00 (s, 3 H), 6.46 (s, 2 H), 7.32 (d, $^{3}J_{HH}$ = 8.8, 2 H), 7.61 (d, $^{3}J_{HH}$ = 8.8, 2 H). ^{13}C NMR (100 MHz, DMSO-d_6) δ: 11.1, 126.7, 128.4, 131.9, 139.2, 141.1. IR (nujol) cm^{-1}: 3365, 3310, 3225, 2970, 2945, 2865, 1640, 1600, 1490, 1430, 1380, 1285, 1135, 1100, 1070, 1015, 970, 840.[3]

7. The submitters do not recommend the use of hydrazone stored more than 7 days as this results in decreased yields.

8. Reagent grade DMSO (99.6% purity) purchased from Sigma Aldrich was used without further purification.

9. Ammonium hydroxide (28% NH_3 in water) was purchased from the Sigma-Aldrich Company.

10. Commercially available copper (I) chloride (98% purity) purchased from Sigma-Aldrich (10 g) was dissolved in 100 mL of 10% HCl, precipitated by dilution with 1000 mL of water and filtered. The solid was washed with water (100 mL), ethanol (100 mL), acetone (100 mL) and ethyl ether (100 mL), and then dried under vacuum (2-3 mm) in a desiccator.

11. Reagent grade CCl_4 (99.5% purity) purchased from Sigma-Aldrich was used without further purification.

12. Intense gas evolution is observed. A characteristic color change from slight-green to blue occurs.

13. Care should be taken to keep the bath temperature at 25°C during rotary evaporation of hexanes because the product is moderately volatile.

14. The main impurity (TLC (hexane:CH_2Cl_2; 2:1): R_f 0.45) was identified as the azine of 1-(4-chlorophenyl)ethanone.

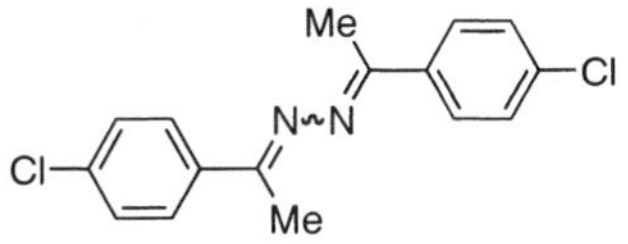

15. ^{1}H NMR (400 MHz, $CDCl_3$) δ: 2.17 (s, 3 H), 7.19 (d, $^3J_{HH}$ = 8.8, 2 H), 7.32 (d, $^3J_{HH}$ = 8.8, 2 H). ^{13}C NMR (100 MHz, $CDCl_3$) δ: 22.9, 117.6, 128.6, 129.2, 133.6, 134.5, 138.4. IR (neat) cm^{-1}: 2930, 2860, 1915, 1650, 1615, 1595, 1490, 1430, 1400, 1380, 1265, 1100, 1070, 1040, 1020, 920, 840. TLC (R_f (hexane) 0.55). Anal calcd for $C_9H_7Cl_3$: C, 48.80; H, 3.18. Found: C, 49.02; H, 3.14.

Safety and Waste Disposal Information

All hazardous materials should be handled and disposed of in accordance with "Prudent Practices in the Laboratory"; National Academy Press; Washington, DC, 1995.

3. Discussion

Development of new catalytic methods for the construction of double carbon-carbon bonds is a goal of modern synthetic chemistry. Olefination of carbonyl compounds, that is, conversion of the carbonyl group into C=C fragment, is a widely used approach for the formation of alkenes. Wittig-type reactions, Peterson and Julia olefination and reductive couplings of carbonyl compounds are the most common olefination methods.[4,5] However, the necessity to use equimolar quantities or a large excess of phosphorus, silicon, sulfur reagents or metals is a significant disadvantage of these techniques.

Table 1

Preparation of Dichloroalkenes

Substrate	Product	Yield
Me–C$_6$H$_4$–C(=NNH$_2$)Me	Me–C$_6$H$_4$–C(Me)=CCl$_2$	65
MeO–C$_6$H$_4$–C(=NNH$_2$)Me	MeO–C$_6$H$_4$–C(Me)=CCl$_2$	70
O_2N–C$_6$H$_4$–C(=NNH$_2$)Me	O_2N–C$_6$H$_4$–C(Me)=CCl$_2$	58
C$_6$H$_5$–C(=NNH$_2$)Et	C$_6$H$_5$–C(Et)=CCl$_2$	57

The present procedure describes a convenient preparation of dichloroalkenes from aryl alkyl ketones based on a novel catalytic olefination reaction (COR) of carbonyl compounds.[6] Some representative examples of dichloromethylenation of ketones are compiled in Table 1.

Recently we found that *N*-unsubstituted hydrazones of aldehydes and ketones could be transformed into alkenes when treated with polyhalogenated alkanes in the presence of copper (I) chloride. This method has already found applications in the synthesis of dihaloalkenes,[6] monohaloalkenes,[7] fluorinated olefins,[8] and a range of alkenes containing functional groups[9] from the corresponding carbonyl precursors (Figure 1). Mild reaction conditions and operational simplicity are important features of this catalytic olefination procedure.

Figure 1

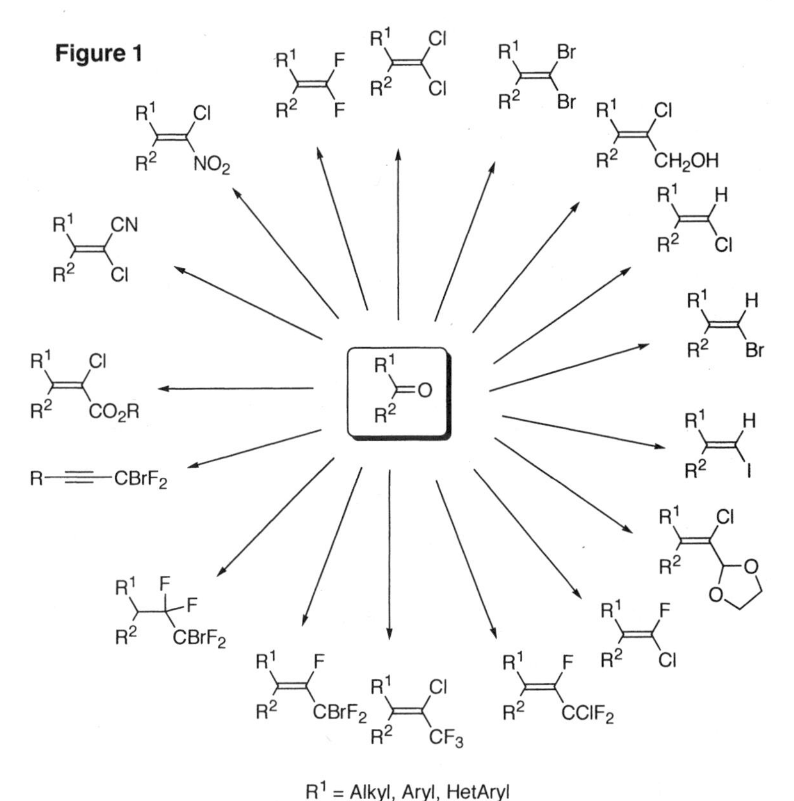

R^1 = Alkyl, Aryl, HetAryl
R^2 = H, Alkyl

1. Moscow State University, Department of Chemistry, Leninskie Gory, Moscow 119992, Russia. E-mail: nen@acylium.chem.msu.ru
2. Institute of Problems of Chemical Physics, Chernogolovka, Moscow region, 142432, Russia
3. Newkome, G. R.; Fishel, D. L. *J. Org. Chem.* **1966**, *31*, 677.
4. Lawrence, N. L. In *Preparation of Alkenes*; Williams, J. M. J., Ed.; Oxford University Press: Oxford, **1996.**
5. Kelly, S. E. In *Comprehensive Organic Synthesis*, eds. Trost, B. M.; Fleming, I. Ed.; Pergamon Press, Oxford, **1992**, pp. 730 – 810.
6. (a) Shastin, A. V.; Korotchenko, V. N.; Nenajdenko, V. G.; Balenkova, E. S. *Tetrahedron* **2000**, *56*, 6557; (b) Shastin, A. V.; Korotchenko, V. N.; Nenajdenko, V. G.; Balenkova, E. S. *Synthesis* **2001**, 2081; (c) Korotchenko, V. N.; Shastin, A. V.; Nenajdenko, V. G.; Balenkova, E. S. *J. Chem. Soc. Perkin Trans. 1* **2002**, 883; (d) Korotchenko, V. N.;

Shastin, A. V.; Nenajdenko, V. G.; Balenkova, E. S. *Org. Biomol. Chem.* **2003**, *1*, 1906.

7. (a) Shastin, A. V.; Korotchenko, V. N.; Nenajdenko, V. G.; Balenkova, E. S. *Russ. Chem. Bull.* **2001**, *50*, 1401; (b) Shastin, A. V.; Korotchenko, V. N.; Varseev G. N.; Nenajdenko, V. G.; Balenkova, E. S. *Russ. J. Org. Chem.* **2003**, *39*, 403.
8. (a) Korotchenko, V. N.; Shastin, A. V.; Nenajdenko, V. G.; Balenkova, E. S. *Tetrahedron* **2001**, *57*, 7519; (b) Nenajdenko, V. G.; Shastin, A. V.; Korotchenko, V. N.; Varseev G. N.; Balenkova, E. S. *Eur. J. Org. Chem.* **2003**, 302; (c) Nenajdenko, V. G.; Varseev G. N.; Korotchenko, V. N.; Shastin, A. V.; Balenkova, E. S. *J. Fluorine Chem.* **2003**, *124*, 115.
9. (a) Nenajdenko, V. G.; Lenkova O. N.; Shastin, A. V.; Balenkova, E. S. *Synthesis* **2004**, 573; (b) Nenajdenko, V. G.; Shastin, A. V.; Golubinskii I. V.; Lenkova O. N.; Balenkova, E. S. *Russ. Chem. Bull.* **2004**, *1*, 218.

Appendix
Chemical Abstracts Nomenclature (Registry Number)

Hydrazine hydrate: Hydrazine, monohydrate; (7803-57-8)

1-(4-Chlorophenyl)ethanone; (99-91-2)

1-(4-Chlorophenyl)ethanone hydrazone; (40137-41-5)

Copper (I) chloride: Copper chloride; (7758-89-6)

Carbon tetrachloride: Tetrachloromethane; (56-23-5)

1,4-DIOXENE

(2,3-Dihydro-1,4-dioxin)

$Pb(OAc)_4$, hν, Δ; FVP, 425 °C

Submitted by Matthew M. Kreilein, James C. Eppich, and Leo A. Paquette.[1]
Checked by Christopher P. Davie and Rick L. Danheiser.

1. Procedure

A. *2-Acetoxy-1,4-dioxane.* An oven-dried, 500-mL, three-necked, round-bottomed flask is equipped with a magnetic stir-bar, a reflux condenser fitted with an argon inlet, a thermometer, and a 20-cm length of black rubber tubing (1/8 in thick, 1 in diameter) attached to a dry 125-mL Erlenmeyer flask wrapped in aluminum foil and containing 91.9 g (207 mmol) of lead tetraacetate (Note 1). The flask is charged with 200 mL (2.35 mol) of dioxane (Note 2) and heated with stirring at 80°C with a "Power Light" 500 W lamp (Note 3). The lead tetraacetate is added portionwise over approximately 30 min while the temperature is increased to the reflux point. A cloudy white or tan solution develops. The reaction mixture is heated at reflux for 1 hr (Note 4) and then allowed to cool to rt. Saturated $NaHCO_3$ solution (200 mL) is carefully introduced, and the resulting mixture is filtered through a pad of Celite. The filtrate is extracted with three 100-mL portions of CH_2Cl_2, and the combined organic fractions are washed with 100 mL of saturated $NaHCO_3$ solution, dried over Na_2SO_4, filtered, and concentrated by rotary evaporation (20 mmHg) at room temperature. The resulting oil (27.3 g) is fractionally distilled at reduced pressure (Note 5) to remove dioxane and deliver 22.0 g (73%) of 2-acetoxy-1,4-dioxane as a colorless liquid, bp 54-55°C (0.4 mmHg) (Note 6).

B. *1,4-Dioxene.* A pyrolysis apparatus is assembled as shown in the photographs below. A 50-mL, two-necked, round-bottomed flask containing 18.0-18.1 g (0.123-0.124 mol) of 2-acetoxy-1,4-dioxane is attached to a quartz tube (Note 7) packed with coarse quartz chips (ca. 1/4") and heated at 425°C in a horizontal furnace (Notes 8, 9). The tube is attached to a U-tube that is charged with ca. 11 g of NaOH pellets. A Dewar (cold finger) condenser with a 105° angled side joint is attached to the other end of the U-tube and filled with dry ice/acetone (–78°C). A vacuum trap, cooled in liquid nitrogen, is positioned between the cold finger and the vacuum source. A gentle flow of N_2 (constricted through a needle valve) is initiated at the side neck of the flask and a gentle vacuum (ca. 265 mmHg, controlled by a digital vacuum regulator) is applied via the angled side joint of the Dewar condenser (Note 10).

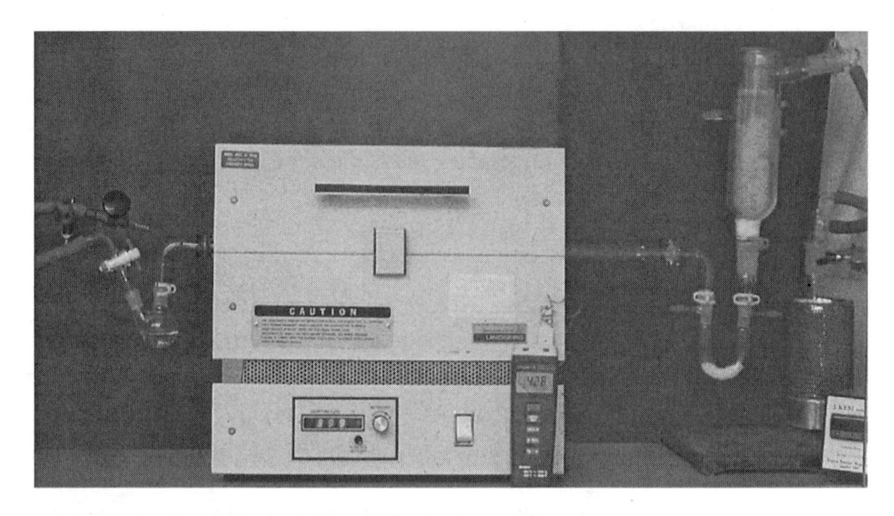

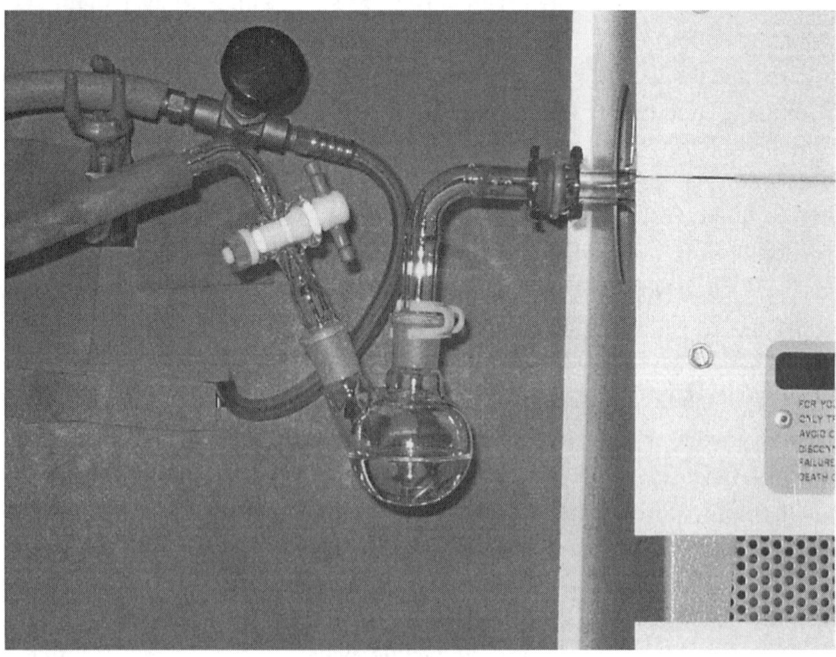

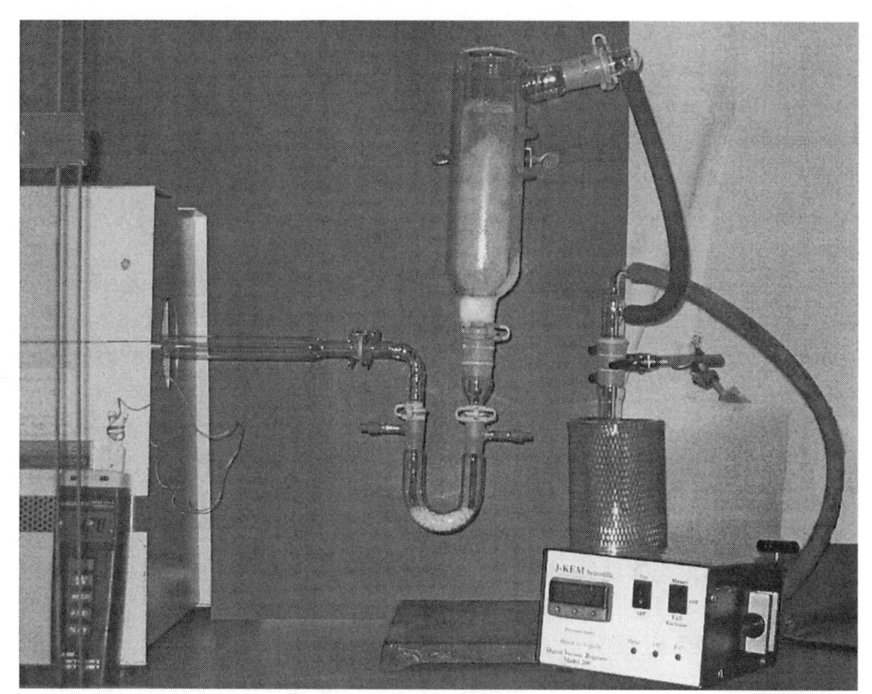

The 2-acetoxy-1,4-dioxane is introduced into the vapor phase by appropriate application of a modest amount of heat from a commercial heat gun (Note 11). After the introduction of the acetate is completed (25 min), the Dewar condenser is mounted above a 25-mL, round-bottomed flask and a mixture of 1,4-dioxene and acetic acid is collected therein following warming of the cold finger bath. The liquids contained in the U-tube (Note 12) and liquid nitrogen trap are combined with the material collected from the cold finger to afford 9.4-9.8 g of a colorless to pale yellow liquid (Note 13). Sodium hydroxide pellets (4.0 g) are added, and the resulting gel-like mixture is allowed to stand for 1 h and then distilled at atmospheric pressure (Note 14). The distillate of water and 1,4-dioxene is placed in a –18°C freezer overnight and 1,4-dioxene is obtained by decantation from the ice crystals to furnish 4.7-6.5 g (44-61%) of 1,4-dioxene as a colorless liquid (Notes 15, 16).

2. Notes

1. Lead tetraacetate was obtained from Acros Organics (95% purity, stabilized with ca. 4% acetic acid) and used as received. If lightly colored, the reagent was used directly. If dark-brown in color, the lead tetraacetate was recrystallized from acetic acid according to the procedure described in

Perrin, D. D.; Armarego, W. L. F.; Perrin, D. R. *Purification of Laboratory Chemicals,* 2nd ed.; Pergamon Press: Oxford, 1980; pp 497-498.

2. Dioxane (anhydrous, 99.8%) was purchased from Alfa Aesar and distilled from calcium hydride under argon before use.

3. The lamp was an inexpensive model purchased from a discount distributor (Sam's Club). The manufacturer was The Designer's Edge, 11730 N.E. 12th St., Bellevue, WA 98006.

4. In the event that gas evolution (as evidenced by briefly turning off the Ar source and watching the bubbler) persists after the reaction mixture is heated at reflux for 1 h, refluxing is continued until gas evolution ceases.

5. A short path distillation head attached to a receiver cow was employed with the receiving flask cooled in a dry ice/acetone bath.

6. In different runs, the yield ranged from 68-73%. The product exhibits the following spectroscopic properties: IR (film): 2976, 2860, 1748, 1454, 1374, 1225 cm^{-1}; ^{1}H NMR (500 MHz, $CDCl_3$) δ: 2.16 (s, 3 H), 3.64 (app dt, J = 11.8, 2.7 Hz, 1 H), 3.71-3.82 (m, 4 H), 4.09-4.17 (m, 1 H), 5.85 (t, J = 2.1 Hz, 1 H); ^{13}C NMR (125 MHz, $CDCl_3$) δ: 21.3, 61.9, 66.3, 67.9, 89.4, 170.0; HRMS (ESI) Calcd for $C_6H_{10}O_4$ $[M + Na]^+$: 169.0471; Found: 169.0478.

7. The dimensions of the quartz tube used by the checkers were as follows: overall length, 590 mm; length of tube between insulation inside oven, 308 mm; outside diameter, 19 mm. The tube used by the submitters had the following dimensions: overall length, 490 mm; length of tube inside oven between insulation, 280 mm; inside diameter, 13 mm.

8. Any type of furnace may be employed, although "hot spots" should be avoided if possible. Should the bore of the oven aperture be significantly larger than the diameter of the quartz tube, the ends of the tube may be wrapped with glass wool tape to fit. The checkers used an oven manufactured by Lindberg (a unit of General Signal), Watertown, WI 53094. The submitters used an oven manufactured by Hevi Duty Electric Company, Milwaukee, WI. The checkers monitored the oven temperature using an Omegaette® Model HH308 digital thermometer equipped with a thermocouple probe that was inserted into the center of the oven. The submitters monitored the oven temperature using a Keithley Model 871 digital thermometer equipped with a ceramic temperature probe that was inserted into the center of the oven.

9. The quartz tube was preheated at 425°C for ≥1h before the 2-acetoxy-1,4-dioxane was introduced.

10. The pyrolysis setup used by the submitters did not include a liquid nitrogen trap. Their U-tube contained ca. 5 g of NaOH, and they controlled the flow of nitrogen with a fine capillary.

11. The checkers used a heat gun with a 260-399°C range (14 amp) manufactured by the Master Appliance Corporation, Racine, WI 53403. The submitters used a heat gun made by the same manufacturer with a 149-260°C range (12 amp).

12. In one run, the checkers found that gently heating the U-tube with a heat gun (after the apparatus had been disassembled) allowed more liquid to be decanted from the U-tube.

13. ^{1}H NMR analysis indicated that this material is a ca. 75:25 mixture of 1,4-dioxene and AcOH.

14. A short path distillation head attached to a 25-mL, round-bottomed receiver flask were used. The mixture of water and 1,4-dioxene distilled at 58-94°C (760 mmHg).

15. The submitters report that they obtained 10-20% pure dioxene from the Dewar condenser cold finger and additional product by distillation of the liquid decanted from the U-tube; total yield: 65%.

16. The product exhibits the following spectroscopic properties: IR (film): 3099, 2982, 2933, 2879, 2023, 1654, 1458, 1395, 1268, 1128, 1067, 954, 898, 739 cm^{-1}; ^{1}H NMR (500 MHz, $CDCl_3$) δ: 4.07 (s, 4 H), 5.96 (s, 2 H); ^{13}C NMR (125 MHz, $CDCl_3$) δ: 64.9, 127.2; Anal. Calcd for $C_4H_6O_2$: C, 55.81; H, 7.02. Found: C, 55.69; H, 7.01.

Safety and Waste Disposal Information

All hazardous materials should be handled and disposed of in accordance with "Prudent Practices in the Laboratory"; National Academy Press; Washington, DC, 1995.

3. Discussion

The 1,4-dioxene molecule (**1**) has attracted attention for a number of years since its initial preparation by Summerbell and Bauer in 1935.[2] The symmetrical nature of its double bond and the *cis* orientation of the two oxygen atoms are features not commonly resident in other structural contexts. One consequence is the lowering of the first ionization potential of **1** to 8.43 eV relative to the corresponding value in dihydropyran (8.84 eV).[3] As a result, this heterocyclic building block undergoes successful [2+2] photocycloaddition to conjugated enones,[4] 1,2-diketones,[5] and benzene.[6]

The facility with which **1** enters into the Paterno-Büchi reaction has also been documented.[7] Comparable interest has surrounded the involvement of 1,4-dioxene in thermal inverse electron demand Diels-Alder processes,[8,9] trapping with molybdenum and chromium carbene complexes,[10] [2+2] ketene cycloadditions,[11] and cyclopropanations with diazo compounds.[12]

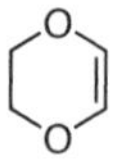

1

The enol ether constitution of **1** has caused it to be regarded as a protecting group for alcohols.[13] Beyond this, conversion to 2-dioxenyllithium (**2**) can be efficiently accomplished by exposure to *tert*-butyllithium in THF at low temperature.[14] This organometallic intermediate has been broadly exploited by Fétizon and Hanna in synthesis,[15] and the derived stannane **3** can be smoothly acylated with acyl chlorides under conditions of palladium catalysis.[16] Added scope is provided by the higher order cuprate **4** whose reactivity is well suited to electrophilic capture.[17]

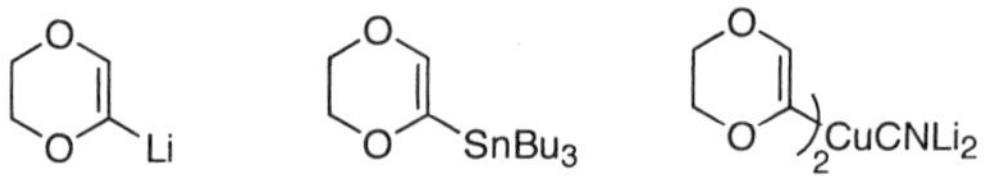

Li SnBu$_3$ $)_2$CuCNLi$_2$

2 **3** **4**

Despite the considerable promise of 1,4-dioxene in organic synthesis, only three preparative routes to **1** have been reported. More astonishing yet was the unsuitability of all three routes for the laboratory-scale acquisition of reasonable amounts of *pure* reagent (Scheme 1). The original pathway proceeds via the photochlorination of *p*-dioxane to the 2,3-dichloro derivative **5** (70%) followed by reductive dehalogenation with magnesium and iodine (49%).[2] Entry has also been made from diethylene glycol (**6**), heating of which with a copper chromite catalyst and $KHSO_4$ in the liquid phase proceeds with oxidation and cyclodehydration (67%).[18] The complication here is the co-production of 2-*p*-dioxanone, a side reaction highly dependent on the proportion of $KHSO_4$ present. The third strategy involves the photoaddition of phenanthrenequinone to *p*-dioxane and thermal activation of the resulting **7** at 230-250°C to liberate **1**.[19]

The two-step process described here results in the clean formation of 1,4-dioxene free of contaminants and is therefore expected to find serviceable application in the synthesis of this useful heterocyclic intermediate. In this more practical and convenient route, advantage is taken of the rarely exploited capability[20,21] of lead tetraacetate to engage in the acetoxylation of C-H bonds positioned at benzylic[22] and allylic sites,[23] as

Scheme 1

5 → 1: Mg, I_2

6 → 1: copper chromite, $KHSO_4$, 230-250 °C

7 → 1: 230-250 °C

well as adjacent to ethereal oxygen centers.[24] In the specific case of *p*-dioxane, the eight available C-H bonds are equivalent by virtue of symmetry, thus simplifying matters considerably. The result of irradiating a refluxing solution of lead tetraacetate in dioxane with an inexpensive commercial 500 W light source for ca. 1 h and subsequent fractional distillation is to provide the 2-acetoxy derivative in 68-73% yield. The boiling point of this colorless liquid is sufficiently higher than that of the starting material to permit its isolation in a pure state. The flash vacuum pyrolysis step results in the smooth thermal extrusion of acetic acid to deliver **1**. When this experiment is conducted in that manner where no acetate remains unreacted, the yield of volatile 1,4-dioxene is 44-61%.

1. Department of Chemistry, The Ohio State University, Columbus, Ohio 43210-1106.
2. Summerbell, R. K.; Bauer, L. N. *J. Am. Chem. Soc.* **1935**, 57, 2364. See also Summerbell, R. K.; Umhoefer, R. R. *J. Am. Chem. Soc.* **1939**, *61*, 3016.
3. Bloch, M.; Brogli, F.; Heilbronner, E.; Jones, T. B.; Prinzbach. H.; Schweikert, O. *Helv. Chim. Acta* **1978**, *61*, 1388.
4. (a) Bernassau, J. M.; Bouillot, A.; Fétizon, M.; Hanna, I.; Maia, E. R.; Prangé, T. *J. Org. Chem.* **1987**, *52*, 1993. (b) Blechert, S.; Jansen, R.; Velder, J. *Tetrahedron* **1994**, *50*, 9649.
5. Horspool, W. M.; Khandelwal, G. D. *Chem. Commun.* **1967**, 1203.
6. (a) Atkins, R. J.; Fray, G. I.; Drew, M. G. B.; Gilbert, A.; Taylor, G. N. *Tetrahedron Lett.* **1978**, *19*, 2945. (b) Mattay, J.; Leismann, H.; Scharf, H. D. *Chem. Ber.* **1979**, *112*, 577. (c) Atkins, R. J.; Fray, G. I.; Gilbert,

A.; bin Samsudin, M. W.; Steward, A. J. K.; Taylor, G. N. *J. Chem. Soc., Perkin Trans. 1* **1979**, 3196. (d) Gilbert, A.; Taylor, G. N.; bin Samsudin, M. W. *J. Chem. Soc., Perkin Trans. 1* **1980**, 869. (e) Gilbert, A. *Pure Appl. Chem.* **1980**, *52*, 2669.

7. (a) Lazear, N. R.; Schauble, J. H. *J. Org. Chem.* **1974**, *39*, 2069. (b) Araki, Y.; Nagasawa, J.; Ishido, Y. *J. Chem. Soc., Perkin Trans. 1* **1981**, 12. (c) Freilich, S. C.; Peters, K. S. *J. Am. Chem. Soc.* **1981**, *103*, 6255. (d) Adam, W.; Kliem, U.; Lucchini, V. *Tetrahedron Lett.* **1986**, *27*, 2953. (e) Buschmann, H.; Hoffmann, N.; Scharf, H. D. *Tetrahedron: Asymmetry* **1991**, *2*, 1429.
8. Thalhammer, F.; Wallfahrer, U.; Sauer, J. *Tetrahedron Lett.* **1990**, *31*, 6851.
9. 1,4-Dioxene is also known to undergo thermal [2+2] cycloadditions to *N*-phenyltriazolinedione [Koerner von Gustorf, E.; White, D. V; Kim, B.; Hess, D.; Leitich, J. *J. Org. Chem.* **1970**, *35*, 1155] and singlet oxygen [Schaap, A. P. *Tetrahedron Lett.* **1971**, *12*, 1757], but not with tetracyanoethylene [Huisgen, R.; Steiner, G. *Tetrahedron Lett.* **1973**, *14*, 3763]. The latter reagent gives rise to a violet charge-transfer complex in $CHCl_3$.
10. Harvey, D. F.; Grenzer, E. M.; Gantzel, P. K. *J. Am. Chem. Soc.* **1994**, *116*, 6719.
11. (a) Huisgen, R.; Feiler, L. A.; Otto, P. *Chem. Ber.* **1969**, *102*, 3444. (b) Fétizon, M.; Hanna, I. *Synthesis* **1990**, 583.
12. (a) Huisgen, R.; Feiler, L. A.; Otto, P. *Chem. Ber.* **1969**, *102*, 3405. (b) Shatzmiller, S.; Neidlein, R. *Justus Liebigs Ann. Chem.* **1977**, 910. (c) Jendralla, H.; Pflaumbaum, W. *Chem. Ber.* **1982**, *115*, 229. (d) Wenkert, E.; Greenberg, R. S.; Raju, M. S. *J. Org. Chem.* **1985**, *50*, 4681.
13. Fétizon, M.; Hanna, I. *Synthesis* **1985**, 806.
14. Saylor, R. W.; Sebastian, J. F. *Synth. Commun.* **1982**, *12*, 579.
15. (a) Fétizon, M.; Hanna, I.; Rens, J. *Tetrahedron Lett.* **1985**, *26*, 3453. (b) Fétizon, M.; Goulaouic, P.; Hanna, I. *Tetrahedron Lett.* **1985**, *26*, 4925. (c) Fétizon, M.; Goulaouic, P.; Hanna, I.; Prangé, T. *J. Org. Chem.* **1988**, *53*, 5672. (d) Fétizon, M.; Goulaouic, P.; Hanna, I. *Heterocycles* **1989**, *28*, 521. (e) Fétizon, M.; Goulaouic, P.; Hanna, I. *J. Chem. Soc., Perkin Trans. 1* **1990**, 1107. (f) Hanna, I. *Tetrahedron Lett.* **1995**, *36*, 889.
16. Blanchot, V.; Fétizon, M.; Hanna, I. *Synthesis* **1990**, 755.
17. (a) Blanchot-Courtois, V.; Hanna, I. *Tetrahedron Lett.* **1992**, *33*, 8087. (b) Boger, D. L.; Zhu, Y. *J. Org. Chem.* **1994**, *59*, 3453.
18. Moss, R. D.; Paige, J. N. *J. Chem. Eng. Data* **1967**, *12*, 452.
19. Rubin, M. B. *Synthesis* **1977**, 266.

20. Butler, R. N. In *Synthetic Reagents;* Pizey, J. S., Ed.; Ellis Horwood Limited: Chichester, 1977; Vol. 3, pp 277-419.
21. Mihailovic, M. Lj.; Cerkovic, Z. In *Encyclopedia of Reagents for Organic Synthesis*; Paquette, L. A., Ed.; Wiley: Chichester, 1995; Vol. 5, pp 2949-2954.
22. (a) Heiba, E. I.; Dessau, R. M.; Koehl, W. J., Jr. *J. Am Chem. Soc.* **1968**, *90*, 1082. (b) Cavill, G. W. K.; Solomon, D. H. *J. Chem. Soc.* **1954**, 3943.
23. Crilley, M. M. L.; Larsen, D. S.; Stoodley, R. J.; Tomé, F. *Tetrahedron Lett.* **1993**, *34*, 3305.
24. Hill, R. K.; Morton, G. H.; Peterson, J. R.; Walsh, J. A.; Paquette, L. A. *J. Org. Chem.* **1985**, *50*, 5528.

Appendix
Chemical Abstracts Nomenclature; (Registry Number)

Lead tetraacetate: Acetic acid, lead(4+) salt; (546-67-8)

Dioxane: 1,4-Dioxane; (123-91-1)

2-Acetoxy-1,4-dioxane: 1,4-Dioxan-2-ol, acetate; (1743-23-3)

1,4-Dioxene: 2,3-Dihydro-1,4-dioxin; (543-75-9)

(*R*)-(+)-3,4-DIMETHYLCYCLOHEX-2-EN-1-ONE

[(*R*)-(+)-3,4-Dimethyl-2-cyclohexen-1-one]

A. 1) O_3, MeOH; 2) $Cu(OAc)_2 \cdot H_2O$, $FeSO_4 \cdot 7H_2O$, MeOH → **1**

B. **1** → MeLi·LiBr, Et_2O → **2**

C. **2** → PCC, CH_2Cl_2 → **3**

Submitted by James D. White, Uwe M. Grether, and Chang-Sun Lee.[1]
Checked by Rick L. Danheiser and Charnsak Thongsornkleeb.

1. Procedure

A. *(R)-(+)-6-Methylcyclohex-2-en-1-one* (**1**). A 500-mL, three-necked, round-bottomed flask is equipped with a mechanical stirrer, thermometer, and a disposable pipette attached via tygon tubing to an ozone generator (Note 1). The flask is charged with (2*R*,5*R*)-(+)-*trans*-dihydrocarvone (10.4 g, 68 mmol) (Note 2) and 200 mL of methanol (Note 3). The solution is cooled to –30°C in a dry ice-acetone bath and ozone is passed through the solution until TLC analysis shows that no dihydrocarvone remains (Note 4). The reaction mixture is flushed with argon for 7 min and then allowed to warm to –20°C. Copper (II) acetate monohydrate (27.2 g, 136 mmol) (Note 5) is added in one portion. The resulting suspension is stirred for 15 min, and ferrous sulfate heptahydrate (22.7 g, 82 mmol) (Note 5) is added in 1 g portions over a 20 min period. The dark green suspension is stirred at –20°C for 7 hr and then at ambient temperature for 3 hr. Water (200 mL) is added and the resulting mixture is extracted with six 125-mL portions of diethyl ether. The combined organic layers are washed with 100 mL of saturated aq sodium hydrogen carbonate (Note 6) and 50 mL of brine, and dried over anhydrous magnesium sulfate. The solvent is removed by rotary evaporation at ambient temperature (50 mmHg) and the residue is purified by column chromatography on 350 g of silica (elution with 6:1 to 4:1 pentane/diethyl ether) (Notes 7 and 8) to give 4.26 g (57%) of (*R*)-(+)-6-methylcyclohex-2-en-1-one as a colorless oil (Notes 9, 10).

B. *(6R)-(+)-1,6-Dimethylcyclohex-2-en-1-ol* (**2**). A flame-dried, 250-mL, three-necked, round-bottomed flask is equipped with a magnetic stir bar, rubber septum, glass stopper, and an argon inlet. The flask is charged with a solution of (*R*)-(+)-6-methylcyclohex-2-en-1-one (4.16 g, 38 mmol) in 70 mL of diethyl ether (Note 3) and cooled at -78°C. A solution of methyllithium-lithium bromide complex in diethyl ether (27.4 mL, 1.5M, 41 mmol) (Note 5) is then added via syringe during 20 min. The cooling bath is removed and the resulting solution is stirred for 3 hr at ambient temperature. After 3 hr, the reaction mixture is cooled at 0°C and 40 mL of water is very slowly added to the yellow solution (Note 11). The aqueous phase is separated and extracted with two 40-mL portions of diethyl ether. The combined organic layers are washed with 40 mL of water, dried over anhydrous magnesium sulfate, and concentrated by rotary evaporation at room temperature (50 mmHg) to afford 4.79 g (100%) of crude (6*R*)-(+)-1,6-dimethylcyclohex-2-en-1-ol (**2**) as a yellow oil which is used in the next step without further purification (Note 12).

C. *(R)-(+)-3,4-Dimethylcyclohex-2-en-1-one* (**3**). A flame-dried, 250-mL, three-necked, round-bottomed flask is equipped with a magnetic stir bar, rubber septum, glass stopper, and an argon inlet. The flask is charged with pyridinium chlorochromate (16.4 g, 76 mmol) (Note 5) and 75 mL of dichloromethane (Note 3). A solution of crude (6*R*)-(+)-1,6-dimethylcyclohex-2-en-1-ol (4.79 g, 38 mmol), prepared as described above, in 25 mL of dichloromethane (Note 3) is transferred into the reaction mixture via cannula over 5 min, and the resulting mixture is stirred at ambient temperature for 3 hr. The reaction mixture is then diluted with 120 mL of diethyl ether (Note 7), the solution is decanted, and the remaining black resinous polymer is thoroughly washed with three 50-mL portions of diethyl ether. The combined dark brown/black ether solution is washed successively with two 100-mL portions of 5% aqueous sodium hydroxide solution, 100 mL of 5% aqueous hydrochloric acid, and two 50-mL portions of saturated aqueous $NaHCO_3$ solution, dried over anhydrous magnesium sulfate, filtered, and concentrated by rotary evaporation at room temperature (50 mmHg) to give 4.42 g of the crude product as a yellow oil. Purification by column chromatography (Note 13) yields 3.71-3.85 g (79-82% overall from **1**) of (*R*)-(+)-3,4-dimethylcyclohex-2-en-1-one (**3**) as a colorless oil (Note 14).[4]

2. Notes

1. The checkers used a Welsbach model T-816 ozone generator and introduced the ozone via a disposable pipette in an open neck of the flask fitted with a reducing adapter. The submitters vented the reaction flask through a trap filled with 40% aqueous sodium bisulfite solution.

2. (+)-Dihydrocarvone was purchased from Aldrich Chemical Company, Inc. as a ca. 80:20 mixture of (2*R*,5*R*)-(+)-*trans*-dihydrocarvone and (2*S*,5*R*)-(+)-*cis*-dihydrocarvone. Prior to use, this material (23 mL, 21 g, 136 mmol) was purified by column chromatography on 450 g of silica (elution with 8:1 to 6:1 hexane/diethyl ether) to afford 13.9 g (91 mmol) of pure (2*R*,5*R*)-(+)-*trans*-dihydrocarvone.

3. The submitters purchased absolute methanol from J. T. Baker, Inc. and used it without further purification. Anhydrous methanol was purchased by the checkers from Mallinckrodt Chemical Company, Inc., and used as received. The submitters freshly distilled dichloromethane from calcium hydride under argon before use and distilled diethyl ether from sodium and benzophenone under argon. Dichloromethane and diethyl ether were purified by the checkers by pressure filtration through activated alumina .

4. TLC analysis is carried out on silica gel (elution with 50% ethyl acetate-hexane, visualization with $KMnO_4$). Dihydrocarvone exhibits R_f = 0.67. The reaction mixture is a pale blue color after ozonolysis is complete.

5. Copper (II) acetate monohydrate (≥ 99.0%) and ferrous sulfate heptahydrate (≥ 99.5%) were purchased from Fluka Chemical Company, Inc., and were used without further purification. Methyllithium, as a complex with lithium bromide, and pyridinium chlorochromate (98%) were purchased from Aldrich Chemical Company, Inc., and were used without further purification.

6. *Caution*: a large amount of carbon dioxide is liberated!

7. The crude product consists of a mixture of α,β and β,γ enone isomers in a ratio of 74:26. The isomers have R_f values of 0.60 and 0.70, respectively (silica gel TLC, elution with 50% ethyl acetate-hexane and visualization with $KMnO_4$). For purification, the product is charged on a column (24 x 50 cm) of 350 g of silica gel (Sorbent Technologies, 32-63 μm) and eluted with 500 mL of 6:1 pentane-ethyl ether. At that point, fraction collection (30-mL fractions) is begun, and elution is continued with 1500 mL of 6:1 pentane-ethyl ether, 1500 mL of 5:1 pentane-ether, and then 1500 mL of 4:1 pentane-ether. The desired product is obtained in fractions 63-116 and the corresponding β,γ-unsaturated isomer is obtained in fractions 38-55 (17% yield). Pentane, purchased from J.T. Baker, and diethyl ether, purchased from Mallinckrodt Chemical Company, Inc. (certified ACS), were used without further purification.

8. The fractions containing the product were concentrated by careful fractional distillation at room temperature and atmospheric pressure through a 10-inch Vigreux column to a volume of ca. 40-50 mL. This step is necessary to avoid loss of the relatively volatile product. Further concentration was conducted by rotary evaporation at room temperature at 50 mmHg.

9. In other runs the product was obtained in 48-63% yield.

10. (*R*)-(+)-6-Methylcyclohex-2-en-1-one has the following physical properties: ^{1}H NMR (500 MHz, $CDCl_3$) δ: 1.15 (d, $J = 7$ Hz, 3 H), 1.70-1.79 (m, 1 H), 2.05-2.11 (m, 1 H), 2.37-2.45 (m, 3 H), 5.99 (td, $J = 10$ Hz, 3 Hz, 1 H), 6.92-6.96 (m, 1 H); ^{13}C NMR (125 MHz, $CDCl_3$) δ: 15.2, 25.7, 31.0, 41.8, 129.6, 149.9, 202.6; IR (neat) cm^{-1}: 3033, 2964, 2932, 1682, 1389, 1215; $[\alpha]_D^{20}$ + 83 (*c* 1.31, MeOH); + 86 (*c* 1.46, $CHCl_3$); lit[2] + 91 (*c* 1.1 $CHCl_3$); lit[3] + 70 (*c* 3.0, $CHCl_3$)); Anal. Calcd for $C_7H_{10}O$: C, 76.33; H, 9.15. Found: C, 76.13; H, 9.13. Enantiomeric excess of the product was determined by the Submitters to be >99:1 (OD column, 92:8 Hexane/*i*-PrOH, 0.6 mL/min).

11. Water is carefully added dropwise from a disposable pipette. Addition of the first ca. 5 mL is accompanied by violent bubbling.

12. (6*R*)-(+)-1,6-Dimethylcyclohex-2-en-1-ol (**2**) is a pungent liquid. ^{1}H NMR analysis of the crude product indicated that it consists of a 67:33 mixture of diastereomers (the configuration of the isomers was not assigned.)

13. The product is charged on a column (19 x 43 cm) of 150 g of silica gel (Sorbent Technologies, 32-63 μm) and eluted with 300 mL of 6:1 pentane-ether. At that point, fraction collection (30-mL fractions) is begun,

and elution is continued with 1200 mL of 6:1 pentane-ether, 500 mL of 4:1 pentane-ether, 500 mL of 2:1 pentane-ether, and then 700 mL of 1:1 pentane-ether. The desired product is obtained in fractions 38-72 and has R_f = 0.47 (silica gel TLC, elution with 50% ethyl acetate-hexane, visualization with *p*-anisaldehyde).

14. The product has the following physical properties: ^{1}H NMR (500 MHz, $CDCl_3$) δ: 1.20 (d, *J* = 7 Hz, 3 H), 1.73-1.80 (m, 1 H), 1.96 (s, 3 H), 2.09-2.15 (m, 1 H), 2.32 (ddd, *J* = 5, 8, 17 Hz, 1 H), 2.38-2.50 (m, 2 H), 5.83 (app s, 1 H); ^{13}C NMR (125 MHz, $CDCl_3$) δ: 17.9, 22.9, 30.5, 34.5, 34.7, 126.5, 166.8, 199.8; IR (neat) cm^{-1}: 3028, 2966, 2877, 1671, 1626, 1378, 1255; $[\alpha]_D^{20}$ + 108 (*c* 1.28, $CHCl_3$; lit[4] + 111 (*c* 1.06)). The purity of the product was determined to be 95% by GC analysis. The enantiomeric excess of the product was determined by the submitters to be 100% (OD column, 95:5 hexane/*i*-PrOH, 0.6 mL/min; OJ Column, 97:3 Hexane/*i*-PrOH, 0.6 mL/min).

Safety and Waste Disposal Information

All hazardous materials should be handled and disposed of in accordance with "Prudent Practices in the Laboratory"; National Academy Press; Washington, DC, 1995.

3. Discussion

The method described for the preparation of (*R*)-(+)-6-methylcyclohex-2-en-1-one (**1**) was first reported by Schreiber[5] and was improved by Solladié and Hutt.[3] An alternative approach to (*R*)-(+)-6-methylcyclohex-2-en-1-one (**1**) commences from cyclohex-2-en-1-one using Enders' methodology.[2,6] This route delivers **1** via a four-step sequence and in an overall yield of 46%, including two steps for attachment and removal of the chiral auxiliary, (*S*)-1-amino-2-methoxymethylpyrrolidine (SAMP).[6] Since (+)-dihydrocarvone is inexpensive and the *trans*- and *cis*- isomers are conveniently separated by either column chromatography on silica[7] or on 100 g scale by fractional crystallization of the corresponding oximes,[8] the route to **1** from (+)-dihydrocarvone is superior for the large scale preparation of (*R*)-(+)-6-methylcyclohex-2-en-1-one (**1**).

The transformation of (*R*)-(+)-6-methylcyclohex-2-en-1-one (**1**) to (*R*)-(+)-3,4-dimethylcyclohex-2-en-1-one (**3**) via (6*R*)-(+)-1,6-dimethylcyclohex-2-en-1-ol (**2**) was first reported by Tokoroyama *et al*[2] and is based upon a procedure by Dauben and Michno.[9] The optical purity of **3** derived from the Tokoroyama synthesis ($[\alpha]_D^{20}$ + 111 (c 1.06, $CHCl_3$))[2] is comparable to that of the present route starting from (+)-dihydrocarvone ($[\alpha]_D^{20}$ + 104 (c 1.04, $CHCl_3$)). Alternatively, **3** can be prepared via a five-step sequence starting from comparatively expensive (*R*)-(+)-pulegone.[4,10-12]

(*R*)-(+)-3,4-Dimethylcyclohex-2-en-1-one (**3**) is a valuable starting material for the asymmetric synthesis of a variety of natural products and their analogues, many of which possess important properties. Examples include irones, e.g. (2*R*,6*S*)-(+)-*cis*-γ-irone (**4**),[7] which are constituents of essential oils that are highly prized ingredients of certain perfumes; the epiaflavinine derivative 3-demethylaflavinine (**5**);[12] the sponge metabolite (+)-agelasimine A (**6**);[8] the clerodane diterpenoid (-)-methyl kolavenate (**7**);[2] and the antiviral antibiotic (-)-ascochlorin (**8**).[4]

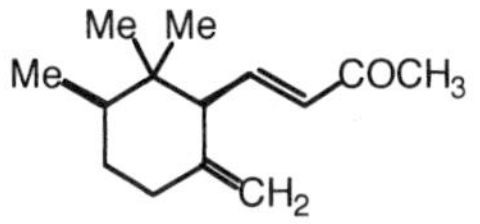

(2*R*,6*S*)-(+)-cis-γ-Irone (**4**)

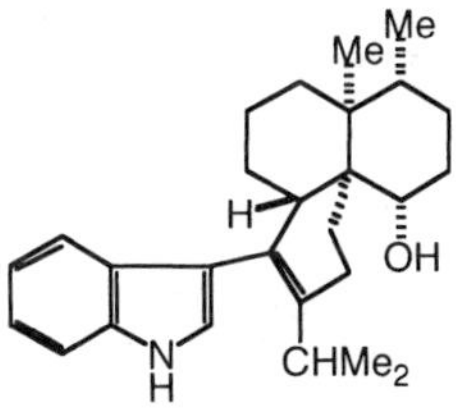

3-Demethylaflavinine (**5**)

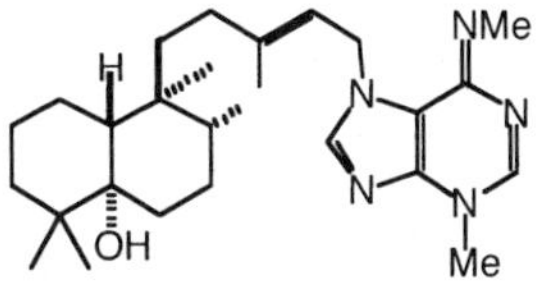

(+)-Agelasimine A (**6**)

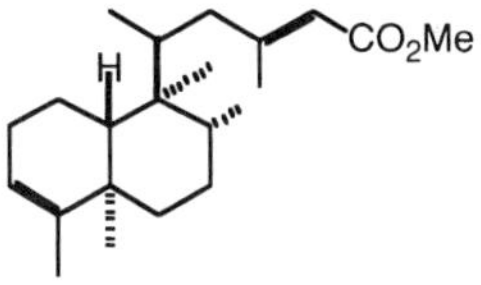

(-)-Methyl Kolavenate (**7**)

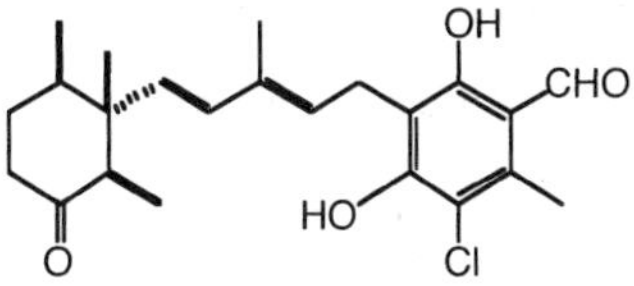

(-)-Ascochlorin (**8**)

1. Department of Chemistry, Oregon State University, Corvallis, Oregon 97331-4003.
2. Iio, H.; Monden, M.; Okada, K.; Tokoroyama, T. *J. Chem. Soc., Chem. Commun.* **1987**, 358-359.
3. Solladié, G.; Hutt, J. *J. Org. Chem.* **1987**, *52*, 3560-3566.

4. Mori, K.; Takechi, S. *Tetrahedron* **1985**, *41*, 3049-3062.
5. Schreiber, S. L. *J. Am. Chem. Soc.* **1980**, *102,* 6163-6165.
6. Enders, D.; Eichenauer, H. *Chem. Ber.* **1979**, *112,* 2933-2960.
7. Laval, G.; Audran, G.; Galano, J. -M.; Monti, H. *J. Org. Chem.* **2000**, *65*, 3551-3554.
8. Ohba, M.; lizuka, K.; Ishibashi, H.; Fujii, T. *Tetrahedron* **1997**, *53*, 16977-16986.
9. Dauben, W. G.; Michno, D. M. *J. Org. Chem.* **1977**, *42,* 682-685.
10. Silvestri, M. G. *J. Org. Chem.* **1983**, *48,* 2419-2420.
11. Danishefsky, S.; Harrison, P.; Silvestri, M.; Segmuller, B. *J. Org. Chem.* **1984**, *49,* 1319-1321.
12. Danishefsky, S.; Chackalamannil, S.; Harrison, P.; Silvestri, M.; Cole, P. *J. Am. Chem. Soc.* **1985**, *107,* 2474-2484.

Appendix
Chemical Abstracts Nomenclature; (Registry Number)

(2*R*,5*R*)-(+)-*trans*-dihydrocarvone: (2*R*,5*R*)-2-Methyl-5-(1-methylethenyl)cyclohexanone; (5948-04-9)

Copper(II) acetate monohydrate: Acetic acid, copper(2+) salt, monohydrate; (6046-93-1)

Ferrous sulfate heptahydrate: Sulfuric acid, iron(2+) salt (1:1), heptahydrate; (7782-63-0)

(*R*)-(+)-6-Methylcyclohex-2-en-1-one: (6*R*)-6-Methyl-2-cyclohexen-1-one; (62392-84-1)

Methyllithium-lithium bromide; (332360-06-2)

(6*R*)-(+)-1,6-Dimethylcyclohex-2-en-1-ol: *trans*-1,6-Dimethyl-2-Cyclohexen-1-ol; (114644-29-0), *cis*-1,6-Dimethyl-2-Cyclohexen-1-ol; (114644-28-9)

Pyridinium chlorochromate; (26299-14-9)

A PRACTICAL AND SAFE PREPARATION OF 3,5-BIS(TRIFLUOROMETHYL)ACETOPHENONE

[1-[3,5-Bis(trifluoromethyl)phenyl]ethanone]

CF_3 / F_3C / Br — *i*-PrMgCl, 0 °C; THF, 1 hr → [CF_3 / F_3C / MgCl, **1**] — Ac_2O, 0 °C; THF, 1 hr → CF_3 / F_3C / O, **2**

Submitted by Johnnie L. Leazer Jr. and Raymond Cvetovich.[1]
Checked by Kevin M. Maloney and Rick L. Danheiser.

1. Procedure[2]

3,5-Bis(trifluoromethyl)acetophenone. A 100-mL, three-necked, round-bottomed flask equipped with a teflon-coated thermocouple probe, inert gas inlet, a 100-mL pressure-equalizing addition funnel fitted with a rubber septum, and a magnetic stir-bar is charged with 3,5-bis(trifluoromethyl)bromobenzene (20.0 g, 11.8 mL, 68.3 mmol) and 35 mL of anhydrous THF (Notes 1, 2, 3). The solution is cooled to -5°C in a ice-salt water bath and a solution of *i*-PrMgCl in THF (37.6 mL, 2M, 75.2 mmol) is added dropwise over 1 hr at a rate such that the internal reaction temperature does not exceed 0°C (Notes 4, 5). Upon completion of the addition, the reaction mixture is stirred for 1 hr at 0 to -10°C.

A 250-mL, three-necked, round-bottomed flask equipped with a teflon-coated magnetic stir bar, a reflux condenser fitted with an inert gas inlet, a glass stopper, and a septum pierced by a teflon-coated thermocouple probe (Note 1) is charged with acetic anhydride (24.4 mL, 258 mmol) (Note 6) and then cooled to -5°C in a ice-salt water bath. The solution of 3,5-bis(trifluoromethyl)phenylmagnesium chloride (at -5°C, prepared as described above) is transferred into the flask containing acetic anhydride via an 18-gauge, 2-ft, double-ended needle at a rate such that the internal reaction temperature does not exceed 0°C (ca. 2 hr) (Note 7). The resulting pale yellow solution is stirred for 30 min at 0°C. The flask is removed from the ice-water bath and 35 mL of deionized water is added dropwise over 1 hr (Note 8). The resulting biphasic mixture is heated at 60°C in an oil bath for 15 min (Note 9). The reaction mixture is then allowed to cool to room temperature. The organic layer is separated (Note 10) and diluted with 35 mL of MTBE (Note 11). The resulting solution is transferred to a 250-mL Erlenmeyer flask and vigorously stirred at 15°C while 2 mL of aqueous 2.5N NaOH is added dropwise so that the pH of the aqueous phase reaches 7.1. The organic phase is separated and washed with two 20-mL portions of saturated aqueous $NaHCO_3$ solution and 20 mL of brine, dried over Na_2SO_4, and concentrated by rotary evaporation (20 mmHg) at room temperature. The residue is transferred to a 50-mL, one-necked, pear-shaped flask. Purification by bulb-to-bulb distillation at 25 mmHg (distillate collected at 122-132°C) affords 15.1-15.2 g (86-87%) of 3,5-bis(trifluoromethyl)-acetophenone as a clear, colorless oil (Note 12).

2. Notes

1. The apparatus was flame-dried and maintained under an atmosphere of argon (checkers) or nitrogen (submitters) throughout the course of the reaction.

2. 3,5-Bis(trifluoromethyl)bromobenzene was purchased from Aldrich Chemical Company, Inc., and used without further purification. Use of 3,5-bis(trifluoromethyl)bromobenzene prepared as described by Leazer et al.[2] gave identical results.

3. The submitters purchased anhydrous THF (HPLC grade) from EM Scientific and used it without further purification. Karl Fischer titration indicated there was 34 μg of water per mL of solvent. The checkers dried THF by pressure filtration under argon through activated alumina.

4. All temperatures refer to internal reaction temperatures.

5. A solution of *i*-PrMgCl (2M in THF) was purchased from Aldrich Chemical Company, Inc., and used as received.

6. Acetic anhydride (certified) was purchased from Fisher Scientific and used without further purification.

7. Addition of acetic anhydride to a freshly prepared solution of **1** gave bis-adduct **3** as the exclusive product. On the other hand, addition of Grignard reagent **1** to an excess of acetic anhydride gave acetophenone **2** in 90% isolated yield. Thus, excess acetic anhydride is required to obtain desired acetophenone

CF3 Ac2O CF3 Ac2O CF3 CF3
Inverse Addition Direct Addition
F3C F3C MgBr F3C CF3
O OAc
2 1 3

8. The reaction mixture warms to room temperature during the addition of water.

9. The 60°C age is necessary in order to destroy residual Ac_2O.

10. The separation of layers is facile. No emulsion problems are encountered. The organic layer is clear and yellow. The aqueous layer is clear and colorless. The reaction mixture turns bright cloudy yellow after addition of 5 mL of water. A gummy residue is observed coating the flask wall after addition of 15 mL of water; however, this residue completely dissolves upon addition of 25 mL of water. Vigorous stirring is used during the addition of water.

11. HPLC grade methyl *tert*-butyl ether (MTBE) was purchased from Fisher Scientific Co. and used without further purification.

12. Product **2** has the following physical properties: ^{1}H NMR ($CDCl_3$, 400 MHz) δ: 2.72 (s, 3 H), 8.09 (s, 1 H), 8.40 (s, 2 H); ^{13}C NMR ($CDCl_3$, 100 MHz) δ: 26.7 (s), 123.2 (q, J = 272.9 Hz), 126.3 (sept, J = 3.6 Hz), 128.3, (q, J = 3.4 Hz), 132.5 (q, J = 34.0 Hz), 138.6 (s), 195.1 (s); ^{19}F NMR ($CDCl_3$, 376 MHz) δ: -63.67; IR (neat) 1705, 1610, 1005, 720 cm^{-1}; Anal. Calcd for $C_{10}H_6F_6O$: C, 46.89; H, 2.36. Found: C, 46.49; H, 2.37.

Safety and Waste Disposal Information

All hazardous materials should be handled and disposed of in accordance with "Prudent Practices in the Laboratory"; National Academy Press; Washington, DC, 1995.

3. Discussion

The trifluoromethylphenyl moiety is a functionality frequently encountered in pharmaceutical drugs,[3] catalysts,[4] and synthetic intermediates.[5] Trifluoromethylphenyl ketones can be classically prepared *via* addition of dimethyl copper lithium to an acid chloride in ethereal solvent;[6] however, the cryogenics make this method unsuitable for large scale synthesis. A more convenient method involves direct acetylation of Grignard reagent **1**. Preparation of the requisite 3,5-bis(trifluoromethyl)-phenyl Grignard reagent is not straightforward, as there are references to detonations of trifluoromethylphenyl Grignard reagents.[7] One such detonation resulted in loss of life.[8] The chemical community is thus in need of a safe and reliable preparation of these valuable reagents.

Our data indicates that detonations associated with formation of trifluoromethylphenyl Grignards using Mg^0 may be attributed to loss of contact with solvent, runaway exothermic side reactions, and potentially, the presence of a highly activated form of Mg. Use of Mg to generate trifluoromethylphenyl Grignard reagents should be avoided if possible. If these reagents are to be generated from Mg, the reaction should be done in THF, and well below reflux, since loss of solvent contact can cause a runaway exothermic reaction.[2] The Knochel procedure should be used whenever possible as it bypasses the factors leading to runaway reactions.[2,9] Knochel's procedure for the low-temperature halogen-magnesium exchange has exhibited no propensity toward runaway reaction. The exchange is rapid (<1h) at cold temperatures (0°C) and suitable for multi-kilogram scale-up.

1. Department of Process Research, Merck Research Laboratories PO Box 2000, Rahway, NJ 07065, USA.
2. A less detailed version of this procedure has been published: Leazer Jr., J. L.; Cvetovich, R.; Tsay, F.-R.; Dolling, U.; Vickery, T.; Bachert, D. *J. Org. Chem.* **2003**, *68*, 3695.
3. (a) Brands, K. M. J.; Payack, J. F.; Rosen, J. D.; Nelson, T. D.; Candelario, M. A.; Huffman, M. A.; Zhao, M. M.; Li, J.; Craig, B.; Song, Z. J.; Tschaen, D. M.; Hansen, K.; Devine, P. N.; Pye, P. J.; Rossen, K.; Dormer, P. G.; Reamer, R.A.; Welch, C. J.; Mathre, D. J.; Tsou, N. N.; McNamara, J. M.; Reider, P. J. *J. Am. Chem. Soc.* **2003**, *125*, 2129. (b) Houlihan, W.J.; Gogerty, J.H.; Ryan, E.A.; and Schmitt, G. *J. Med. Chem.* **1985**, *28*, 28. (c) John T. Welch and Seetha Eswarakrishnan, Fluorine in Bioorganic Chemistry. John Wiley & Sons, p. 246, NY, 1991. (d) Desai, R.C.; Cicala, P.; Meurer, L.C.; Finke, P.E. *Tetrahedron Lett.* **2002**, *43*, 4569. (e) Kuethe, J.T.; Wong, A.; Wu, J.; Davies, I.W.; Dormer, P.G.; Welch, J.W.; Hillier, M.C.; Hughes, D.L.; Reider, P.J. *J. Org. Chem.* **2002**, *67,* 5993. (f) Riachi, N.J.; Arora, P.K.; Sayre, L.M.; and Harik, S.I. *J. Neurochem.* **1988**, 1319.
4. Kaul, F.A.R; Puchta, G.T.; Schneider, H.; Grosche, M.; Mihalios, D.; Herrmann, W.A.; *J. Organomet. Chem.* **2000**, *621*, 184.
5. Pinho, P.; Guijarro, D.; Andersson, P.G.; *Tetrahedron* **1998**, *54*, 7897. (b) Broeke, J.; Deelman, B-J.; Koten, G. *Tetrahedron Lett.* **2001**, *42*, 8085.
6. Posner, G.H.; Whitten, C.E. *Tetrahedron Lett.* **1970**, *11,* 4647.
7. (a) Broeke, J.; Deelman, B-J.; Koten, G. *Tetrahedron. Lett.* **2001**, *42,* 8085. (b) Beck, C.; Park, Y-J.; Crabtree, R. *Chem. Commun.* **1998**, 693. (c) Pinho, P.; Guijarro, D.; Andersson, P. *Tetrahedron* **1998**, *54*, 7897. (d) Kaul, F.; Puchta, G.; Schneider, H.; Grosche, M.; Mihalios, D.; Herrmann, W. *J. Organomet. Chem.* **2001**, *621*, 184. (e) Li, N-S.; Yu, S.;

Kabalka, G. *J. Organomet. Chem.* **1997,** *531,* 101. (f) Doctorvich, F.; Deshpande, A.; Ashby, E. *Tetrahedron* **1994,** *50,* 5945.

8. (a) Appleby, I.C. *Chemistry and Industry*. **1971**, 120. (b) Post Tribune; Gary, Indiana; November 6, 1981.
9. Abarbri, M.; Dehmel, F.; Knochel, P. *Tetrahedron Lett.* **1999**, *40*, 7449.

Appendix
Chemical Abstracts Nomenclature (Registry Number)

3,5-Bis(trifluoromethyl)bromobenzene: 1-Bromo-3,5-bis(trifluoromethyl) benzene; (328-70-1)

3,5-Bis(trifluoromethyl)acetophenone: 1-[3,5-Bis(trifluoromethyl)phenyl] ethanone; (30071-93-3)

Isopropylmagnesium chloride: Magnesium, chloro(1-methylethyl)-; (1068 55-9)

Acetic anhydride: Acetic acid, anhydride; (108-24-7)

ASYMMETRIC ALCOHOLYSIS OF *MESO*-ANHYDRIDES MEDIATED BY ALKALOIDS

(Bicyclo[2.2.1]hept-5-ene-2-*endo*-carboxylic acid, 3-*endo*-benzyloxycarbonyl, (2*R*,3*S*)-)

quinidine (1.1 eq.)
BnOH (3.0 eq.)
toluene, -55 °C, 96 hr

CO_2Bn
CO_2H

Submitted by Carsten Bolm,[1] Iuliana Atodiresei, and Ingo Schiffers.
Checked by Motomu Kanai and Masakatsu Shibasaki.

1. Procedure

(2R,3S)-3-endo-benzyloxycarbonyl bicyclo[2.2.1]hept-5-ene-2-endo-carboxylic acid. A flame-dried 250-mL single-necked, round-bottomed flask equipped with a magnetic stirring bar and charged with quinidine (7.14 g, 22 mmol) (Note 1) and *endo*-bicyclo[2.2.1]hept-5-ene-2,3-dicarboxylic anhydride (3.28 g, 20 mmol) (Note 2) is placed under vacuum for 2 hr (Note 3). The evacuated flask is flushed with argon, charged with 100 mL of dry toluene (Note 4), equipped with a rubber septum and the mixture is cooled to -55°C (Note 5). Benzyl alcohol (6.2 mL, 6.49 g, 60 mmol) (Note 6) is added dropwise via syringe (over a period of 10 min) to the cooled suspension (Note 7), and the reaction mixture is stirred at the indicated temperature for 96 hr (Note 8), during which the solid material gradually dissolves. The resulting clear solution is concentrated *in vacuo* to dryness, and the resulting residue is dissolved in diethyl ether (125 mL). The solution is washed with 2N HCl (3 × 30 mL), and the aqueous layer is back-extracted with ether (5 × 50 mL). The combined organic layers are extracted with a saturated solution of sodium bicarbonate (5 × 75 mL), and the resulting aqueous phase is washed with diethyl ether (1 × 100 mL) in order to remove the traces of benzyl alcohol. The aqueous phase is acidified with 8 N HCl, extracted with CH_2Cl_2 (5 × 100 mL), and the combined organic layers are dried ($MgSO_4$), filtered and concentrated to provide 4.97-5.20 g (91-95%, 97-98% ee) of the benzyl hemi-ester as a white solid (Note 9). The enantiomeric excess of the half-ester was analyzed by chiral HPLC analysis (Note 10). Alternatively, the enantiomeric excess of the benzyl ester[2] could be determined by GC analysis of the corresponding lactone,[3] which was obtained by selective

reduction of the ester group with lithium triethylborohydride ($LiBEt_3H$) followed by acid-catalyzed lactonization (Note 11).[4]

Use of quinine in the ring opening of *endo*-bicyclo[2.2.1]hept-5-ene-2,3-dicarboxylic anhydride provides 5.13 g (94%) of the corresponding enantiomeric benzyl hemi-ester as a white solid (Note 16).

2. Notes

1. Anhydrous quinidine (95%) and quinine (99%) were purchased from Acros Organics and used as supplied.

2. *endo*-Bicyclo[2.2.1]hept-5-ene-2,3-dicarboxylic anhydride (97%) was purchased from Fluka and used as received.

3. High vacuum is used in order to remove traces of moisture.

4. Toluene (Merck, >99%), distilled from sodium benzophenone ketyl radical under argon, was stored over 4Å molecular sieves.

5. An RL 6 CP type cooling machine was used in order to maintain the low temperature.

6. Anhydrous benzyl alcohol (99.8%) was purchased from Sigma-Aldrich and used as supplied.

7. The mixture is stirred for at least 1 hr at the indicated temperature before the benzyl alcohol addition.

8. The reaction has been studied extensively on a 1-mmol scale, and the best asymmetric induction was achieved when the reactions were performed at low temperature. Slightly lower enantioselectivities have been observed when the desymmetrizations were carried out at room temperature.

9. The product has the following characteristics: mp 120°C (racemate), 88-90°C (enantiomer); $[\alpha]_D^{rt}$ = +6.6 (*c* = 1.90, $CHCl_3$); ee = 97-98% (GC-analysis of the lactone: Lipodex E, t_1 = 89.3, t_2 = 89.8 major); ^1H NMR (400 MHz, $CDCl_3$) δ: 1.33 (d, J = 8.6 Hz, 1 H), 1.48 (dt, J = 8.6, 1.8 Hz, 1 H), 3.18 (br s, 1 H), 3.30 (br s, 1 H), 3.31-3.35 (m, 2 H), 4.91 (d, J = 12.6 Hz, 1 H), 5.09 (d, J = 12.6 Hz, 1 H), 6.22 (dd, J = 3.0, 5.7 Hz, 1 H), 6.30 (dd, J = 3.0, 5.7 Hz, 1 H), 7.27-7.36 (m, 5 H); ^{13}C NMR (100 MHz, $CDCl_3$) δ: 46.1, 46.5, 48.1, 48.2, 48.7, 66.3, 128.0, 128.2, 128.4, 134.3, 135.5, 135.8, 172.2, 178.7; IR (KBr) 3065, 3032, 2979, 1740, 1707, 1172 cm^{-1}; EI-MS *m/z* = 272 (M^+, 2), 254 (3), 226 (3), 181 (58), 163 (3), 137 (5), 119 (2), 91 (100), 66 (20). Anal. Calcd for $C_{16}H_{16}O_4$ (272.30): C, 70.57; H, 5.92. Found: C, 70.55; H, 6.01.

10. Conditions: Daicel CHIRALPAK AS-H, eluent = ^{i}PrOH/hexane = 1/1, flow = 0.5 mL/min, detection = 254 nm, retention time = 10 min

(minor isomer using quinidine) and 12 min (major isomer using quinidine).

11. General procedure for the lactone formation. A 25-mL flame-dried Schlenk-flask equipped with a magnetic stirring bar and rubber septum is purged with argon and charged with (2*R*,3*S*)-3-*endo*-benzyloxycarbonyl bicyclo[2.2.1]hept-5-ene-2-*endo*-carboxylic acid (80 mg, 0.29 mmol) in THF (2 mL) (Note 12). The reaction mixture is cooled to 0°C with an ice-bath and 6 eq. of $LiBEt_3H$ (2 mL, 1 M solution in THF) (Note 13) are slowly added. After stirring for 1 hr at room temperature 2 N HCl (5 mL) is slowly added and the mixture is stirred for additional 2 hr. The aqueous phase is extracted with ethyl acetate (3 × 5 mL), and the combined organic phases are dried over $MgSO_4$ and filtered. Evaporation of the solvent yields the corresponding lactone, which is analyzed by GC (Note 14).

In order to recover the alkaloid, the acidic aqueous phase, obtained after the first extraction, is neutralized with Na_2CO_3 and extracted with CH_2Cl_2 (5 × 100 mL). The combined organic phases are dried over $MgSO_4$ and filtered. Evaporation of the solvent yields the alkaloid almost quantitatively (6.93 g, 21.36 mmol, 97%) (Note 15).

12. THF was distilled from sodium benzophenone ketyl radical under argon.

13. $LiBEt_3H$ was purchased from Acros Organics and used as received.

14. Capillary gas chromatograms were obtained using the following column and temperature program: Lipodex E: 2,6-*O*-Dipentyl-3-*O*-butyryl-γ-CD. Column head pressure: 1.0 bar N_2; 100°C (50 min), heating rate 3.0°C/min up to 180°C (60 min). Injector temperature 200°C, detector temperature 250°C.

15. The spectral properties of the recovered alkaloid were in accordance with the data published in the literature.[5]

16. The reaction was performed in a 0.1 M solution (with respect to the anhydride) to give (2*S*,3*R*)-3-*endo*-benzyloxycarbonyl-bicyclo[2.2.1]hept-5-ene-2-*endo*-carboxylic acid with 96% ee (GC-analysis of the lactone: Lipodex E, t_1 = 89.3, major, t_2 = 89.8); $[\alpha]^{rt}_D = -7.4$ (c = 1.00, $CHCl_3$).

Safety and Waste Disposal Information

All hazardous materials should be handled and disposed of in accordance with "Prudent Practices in the Laboratory"; National Academy Press; Washington, DC, 1995.

Table 1. Quinidine-mediated ring opening of various meso-anhydrides with benzyl alcohol.[a]

Entry	Substrate[b]	Product	Yield (%)	ee (%)[c]
1	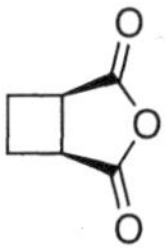 O, O, O	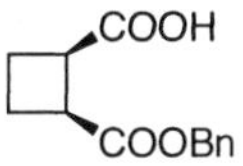 COOH, COOBn	97	96
2	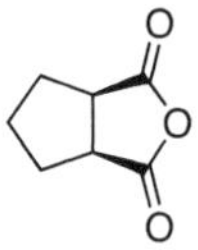 O, O, O	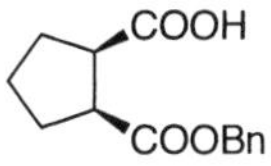 COOH, COOBn	93	97
3	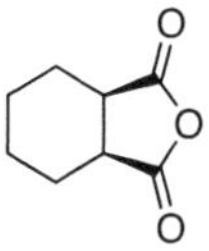 O, O, O	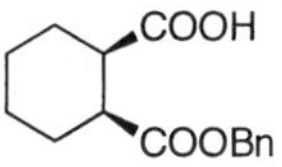 COOH, COOBn	95	97
4	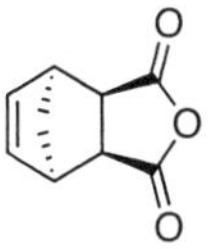 O, O, O	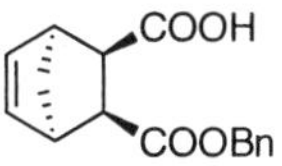 COOH, COOBn	95	97
5[d]	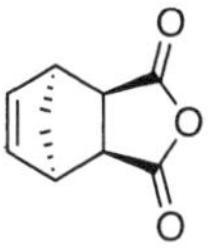 O, O, O	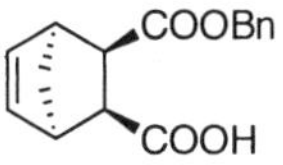 COOBn, COOH	94	96
6	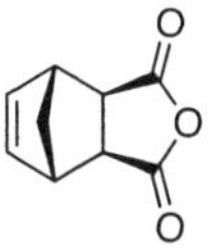 O, O, O	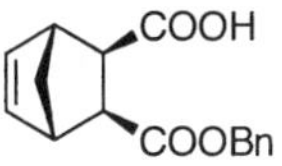 COOH, COOBn	97	96
7	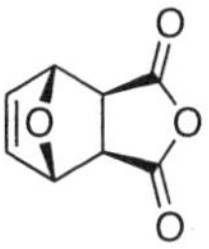 O, O, O, O	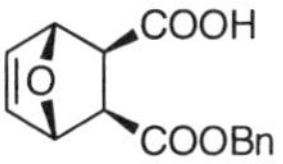 O, COOH, COOBn	89[e]	99

[a] All reactions were performed in toluene, at –55 °C for 96 hr using 1.1 eq. of quinidine and 3 eq. of benzyl alcohol in a 0.2 M solution related to anhydride. All products have been fully characterized.

[b] All meso-anhydrides were prepared by application of literature procedures or were commercially available. For specific procedures, see ref. 10-14.

[c] Determined by GC-analysis of the corresponding lactones using a chiral stationary phase. For retention times, see ref. 3b.

[d] Quinine was used as chiral mediator (0.1 M solution related to anhydride).

[e] After chromatographic purification.

3. Discussion

The procedure described above is an improved version of previously reported alkaloid-mediated asymmetric anhydride openings.[2,3,6,7] Structurally diverse anhydrides have been converted into their corresponding benzyl monoesters with very high enantiomeric excesses and excellent yields. (Table 1) Both enantiomers are available by using either quinidine or quinine as directing additive. An advantage of the present protocol using benzyl alcohol as the nucleophile is that the reactions can be performed using toluene as solvent, avoiding the use of the previously utilized carbon tetrachloride. A simple aqueous work-up permits the isolation of the products in analytically pure form. The synthetic usefulness of the method was demonstrated by the preparation of optically active β-amino acids[2] and unsymmetrical norbornane scaffolds as inducers for hydrogen bond interactions in peptides.[8] In these applications, the benzyl hemi-esters were converted into the corresponding *N*-Cbz-protected β-amino acid benzyl esters by Curtius degradation, which proceeded with neither racemization nor epimerization. Subsequent cleavage of both protecting groups by simple hydrogenation yielded the corresponding free β-amino acids in excellent yields in a single step. Finally, the method is also highly selective for the preparation of the corresponding methyl hemiesters, which have been isolated with up to 99% yield and 99% ee (on a 1-mmol scale). The usefulness of such products for the synthesis of optically active β-amino acids, 1,2-diamines, and polymeric materials has also already been demonstrated.[9]

1. Institut für Organische Chemie der RWTH Aachen, Landoltweg 1, D-52056 Aachen, Germany.

2. Bolm, C.; Schiffers, I.; Atodiresei, I.; Hackenberger, C. P. R. *Tetrahedron: Asymmetry* **2003**, *14*, 3455.
3. (a) Bolm, C.; Gerlach, A.; Dinter, C. L. *Synlett* **1999**, 195. (b) Bolm, C.; Schiffers, I.; Dinter, C. L.; Gerlach, A. *J. Org. Chem.* **2000**, *65*, 6984.
4. Jaeschke, G.; Seebach, D. *J. Org. Chem.* **1998**, *63*, 1190.
5. Raheem, I. T.; Goodman, S. N.; Jacobsen, E. N. *J. Am. Chem. Soc.*, **2004**, *126*, 706.
6. (a) Hiratake, J.; Inagaki, M.; Yamamoto, Y.; Oda, J. *J. Chem. Soc., Perkin Trans. 1* **1987**, 1053. (b) Hiratake, J.; Yamamoto, Y.; Oda, J. *J. Chem. Soc., Chem. Commun.* **1985**, 1717. (c) Aitken, R. A.; Gopal, J. *Tetrahedron: Asymmetry* **1990**, *1*, 517. (d) Aitken, R. A.; Gopal, J.; Hirst, J. A. *J. Chem Soc., Chem. Commun.* **1988**, 632.

7. For recent reviews, see: (a) Chen, Y.; McDaid, P.; Deng, L. *Chem. Rev.* **2003**, *103*, 2965. (b) Tian, S.-K.; Chen, Y.; Hang, J.; Tang, L.; McDaid, P.; Deng, L. *Acc. Chem. Res.* **2004**, *37*, 621.
8. Hackenberger, C. P. R.; Schiffers, I.; Runsink, J.; Bolm, C. *J. Org. Chem.* **2004**, *69*, 739.
9. (a) Bolm, C.; Schiffers, I.; Dinter, C. L.; Defrère, L.; Gerlach, A.; Raabe, G. *Synthesis* **2001**, 1719. (b) Bolm, C.; Dinter, C. L.; Schiffers, I.; Defrère, L. *Synlett* **2001**, 1875. (c) Bolm, C.; Schiffers, I.; Atodiresei, I.; Ozcubukcu, S.; Raabe, G. *New J. Chem.* **2003**, *27*, 14.
10. *cis*-1,2-Cyclobutanedicarboxylic anhydride was obtained by refluxing *cis*-cyclobutane-1,2-dicarboxylic acid (Fluka, >97%) in trifluoroacetic anhydride (Acros Organics, 99+%) for 16 hr.
11. *cis*-1,2-Cyclopentanedicarboxylic anhydride was prepared in a 3-step synthesis according to a literature procedure. Padwa, A.; Hornbuckle, S. F.; Fryxell, G. E.; Stull, P. D. *J. Org. Chem.* **1989**, *54*, 817.
12. *cis*-1,2-Cyclohexanedicarboxylic anhydride (99%) was purchased from Acros Organics and used as received.
13. *Exo*-bicyclo[2.2.1]hept-5-ene-2,3-dicarboxylic anhydride was prepared according to a literature procedure. Canonne, P.; Belanger, D.; Lemay, G. *J. Org. Chem.* **1982**, *47*, 3953. It is also available from Sigma-Aldrich (95%).
14. *Exo*-7-oxabicyclo[2.2.1]hept-5-ene-2,3-dicarboxylic anhydride was prepared by Diels-Alder reaction of maleic anhydride and furan. It is also available from different commercial suppliers (Fluka, Sigma-Aldrich, Acros Organics, Lancaster Synthesis).

Appendix
Chemical Abstracts Nomenclature; (Registry Number)

Quinidine: Cinchonan-9-ol, 6'-methoxy-, (9*S*)-; (56-54-2)
endo-Bicyclo[2.2.1]hept-5-ene-2,3-dicarboxylic anhydride: 4,7-Methanoisobenzofuran-1,3-dione, 3a,4,7,7a-tetrahydro-, (3a*R*,4*S*,7*R*,7a*S*)-rel-; (129-64-6)
Benzyl alcohol: Benzenemethanol; (100-51-6)
Lithium triethylborohydride: Borate(1-), triethylhydro-, lithium, (T-4)-; (22560-16-3)
Quinine: Cinchonan-9-ol, 6'-methoxy-, (8α,9*R*)-; (130-95-0)

IRIDIUM-CATALYZED C-H BORYLATION OF ARENES AND HETEROARENES: 1-CHLORO-3-IODO-5-(4,4,5,5-TETRAMETHYL-1,3,2-DIOXABOROLAN-2-YL)BENZENE AND 2-(4,4,5,5,-TETRAMETHYL-1,3,2-DIOXABOROLAN-2-YL)INDOLE

[2-(3-Chloro-5-iodophenyl)-4,4,5,5-tetramethyl-1,3,2-dioxaborolane]
[2-(4,4,5,5-Tetramethyl-1,3,2-dioxaborolan-2-yl)-1*H*-indole]

A. 2 + 1/2[Ir(OMe)(COD)]$_2$-dtbpy, hexane, 50°C → 2

B. + 1/2[Ir(OMe)(COD)]$_2$-dtbpy, hexane, 25°C →

dtbpy = (H$_3$C)$_3$C ... C(CH$_3$)$_3$

Submitted by Tatsuo Ishiyama, Jun Takagi, Yusuke Nobuta, and Norio Miyaura.[1]
Checked by Yili Shi, Daniel J. Weix and Jonathan A. Ellman.

1. Procedure

Caution! The reactions produce hydrogen gas and should be conducted in a well-ventilated hood.

A. *1-Chloro-3-iodo-5-(4,4,5,5-tetramethyl-1,3,2-dioxaborolan-2-yl)benzene.* A 50-mL, two-necked, round-bottomed flask is fitted with a magnetic stirring bar, a rubber septum, and a condenser to which a nitrogen inlet and an oil bubbler are attached, and flushed with nitrogen (Note 1). The septum is removed and the flask is charged with bis(η^4-1,5-cyclooctadiene)-di-μ-methoxydiiridium(l) ([Ir(OMe)(COD)]$_2$) (33 mg, 0.050 mmol) (Note 2), 4,4'-di-*tert*-butyl-2,2'-bipyridine (dtbpy) (27 mg, 0.10

mmol) (Note 3), and bis(pinacolato)diboron (2.67 g, 10.5 mmol) (Note 4). The septum is again placed on the flask, and the flask is purged with nitrogen for 1 min. Hexane (30 mL) (Note 5) is added by syringe, and the flask is immersed in an oil bath that is maintained at 50°C. The mixture is stirred for 10 min to give a dark red solution. The flask is charged with 1-chloro-3-iodobenzene (4.75 g, 19.9 mmol) by syringe (Note 6), and then the resulting dark red solution is stirred at 50°C for 6 hr (Note 7). The mixture is removed from the oil bath, allowed to cool to room temperature, and poured into a separatory funnel. The reaction flask is rinsed with hexane (2 x 10 mL). The rinses and water (30 mL) are added to the separatory funnel, the funnel is shaken, the layers are separated, and the organic extracts are dried over anhydrous magnesium sulfate. The drying agent is removed by filtration and is washed with hexane (3 x 10 mL), and the filtrate is concentrated on a rotary evaporator to give a dark brown oil. The oil is distilled under reduced pressure (Note 8) to afford 6.11-6.20 g (84-86%) of 1-chloro-3-iodo-5-(4,4,5,5-tetramethyl-1,3,2-dioxaborolan-2-yl)benzene as a white solid, mp 58.6-60.7°C (Note 9).

B. 2-(4,4,5,5-Tetramethyl-1,3,2-dioxaborolan-2-yl)indole. A 50-mL, two-necked, round-bottomed flask is fitted with a magnetic stirring bar, a rubber septum, and a condenser to which a nitrogen inlet and an oil bubbler are attached, and flushed with nitrogen (Note 1). The septum is removed and the flask is charged with bis(η^4-1,5-cyclooctadiene)-di-μ-methoxydiiridium(I) ([Ir(OMe) (COD)]$_2$) (50 mg, 0.075 mmol) (Note 2) and 4,4'-di-*tert*-butyl-2,2'-bipyridine (dtbpy) (40 mg, 0.15 mmol) (Note 3). The septum is again placed on the flask, and the flask is purged with nitrogen for 1 min. Hexane (30 mL) (Note 5) and pinacolborane (4.79 mL, 33.0 mmol) (Note 10) are added by syringe, and the flask is immersed in an oil bath that is maintained at 25°C. The mixture is stirred for 10 min. to give a dark red solution. The flask is charged with indole (3.51 g, 30.0 mmol) (Note 11), and then the resulting dark red suspension is stirred at 25°C for 4 hr. The mixture is removed from the oil bath, allowed to cool to room temperature, and poured into a separatory funnel. The reaction flask is rinsed with ether (2 x 20 mL). The rinses and water (30 mL) are added to the separatory funnel, the funnel is shaken, the layers are separated, and the organic extracts are dried over anhydrous magnesium sulfate. The drying agent is removed by filtration and is washed with ether (3 x 10 mL), and the filtrate is concentrated on a rotary evaporator to give a dark brown oil. The oil is distilled under reduced pressure (Note 12) to afford 5.33-5.40 g (73-74%) of 2-(4,4,5,5-tetramethyl-1,3,2-dioxaborolan-2-yl)indole as a colorless viscous oil, bp 149-153°C/0.15 mm (Note 13).

2. Notes

1. All glassware is pre-dried in an oven at 120°C for 1 hr, assembled while hot, and allowed to cool under a stream of nitrogen.

2. $[Ir(OMe)(COD)]_2$ is prepared from $[IrCl(COD)]_2$ by the reported procedure.[2] The complex can be handled in air, and stored in a tightly closed bottle and in a freezer. Checkers stored the complex in a vial in a nitrogen-filled glove-bag at rt.

3. 4,4'-Di-*tert*-butyl-2,2'-bipyridine (98%) was purchased from Aldrich Chemical Company, Inc. and used without further purification.

4. Bis(pinacolato)diboron is prepared according to the literature procedure,[3] and dried under reduced pressure (0.1 mm) at room temperature for 16 hr. prior to use. The reagent is air-stable and commercially available. The checkers purchased bis(pinacolato)diboron (98%) from Aldrich and used it without further purification.

5. Hexane is distilled from lithium aluminum hydride under nitrogen before use. The checkers purified hexanes by passage through a column of activated alumina (type A2, size 12 × 32, Purifry Co.) followed by nitrogen sparge. The hexanes were stored over MS 3Å in a nitrogen-filled glove-box.

6. 1-Chloro-3-iodobenzene was purchased from Tokyo Kasei Kogyo Co., Ltd., and distilled from molecular sieves 4Å under nitrogen before use. The checkers purchased 1-chloro-3-iodobenzene (98%) from Aldrich.

7. The reaction using 1.5 mol% of the catalyst is completed within 4 hr at room temperature, but the decreasing the amount of catalyst to 0.5 mol% sufficiently slows the rate of the reaction that heating to 50°C is required to complete the reaction in a reasonable amount of time.

8. bp 145-148°C/0.3 mm; 1-Chloro-3-iodobenzene (28 mg) that is unreacted or is generated by protodeboration of the product is also obtained, bp 60°C/0.05 mm. The checkers found that either simple distillation with a short-path distillation head or Kugelrohr distillation can be used. To avoid contamination of the product with lower boiling materials (e.g. starting materials or pinacol), the mixture should be heated until 1-chloro-3-iodobenzene and pinacol have distilled. The apparatus is then allowed to cool, is disassembled, and is cleaned before the product is distilled over.

9. Gas chromatographic analysis of the product (Hitachi G-3500, OV-101 on UniportB, a glass column, 3 mm x 2 m) shows that the chemical purity is 97-99%. The spectral data are as follows: ^{1}H NMR (400 MHz, $CDCl_3$) δ: 1.30 (s, 12 H), 7.72 (d, 1 H, $J = 2.0$), 7.78 (t, 1 H, $J = 1.8$), 7.99 (t, 1 H, $J = 0.8$); ^{13}C NMR (100 MHz, $CDCl_3$) δ: 24.8, 84.5, 94.2, 133.7, 134.7,

139.4, 141.4; ^{11}B NMR (128.3 MHz, $CDCl_3$): δ 29.58 (BF_3•OEt_2 as external reference, δ 0.00); HRMS calcd for $C_{12}H_{15}BClIO_2$[M^+], 363.9899, found 363.9890.

10. Pinacolborane (97%) was purchased from Aldrich Chemical Company, Inc., and used without further purification. Pinacolborane is very sensitive to moisture and was troublesome to store for any period of time. The checkers found it could be stored under nitrogen at -30°C in a sealable tube.

11. Indole was purchased from Tokyo Kasei Kogyo Co., Ltd., and dried under reduced pressure (0.1 mm) at room temperature for 16 hr prior to use. The checkers purchased indole (99%+) from Aldrich.

12. The checkers accomplished the distillation with the apparatus below.

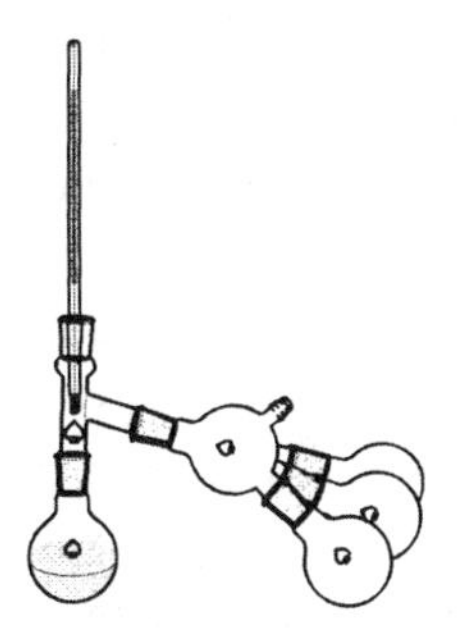

It was found that distillation with a traditional short-path led to clogging of the apparatus by both remaining indole starting material and the highly viscous product. Indole (29 mg) that is unreacted or resulted by protodeboration of the desired product is also obtained, bp 120°C/0.7 mm. To avoid contamination of the product with lower boiling materials (e.g. starting materials or pinacol), the mixture should be heated until indole and pinacol have distilled over. The apparatus is then allowed to cool, is disassembled, and is cleaned before the product is distilled over.

13. Gas chromatographic analysis of the product (Hitachi G-3500, OV-101 on Uniport B, a glass column, 3 mm x 2 m) shows that the chemical purity is 96-99%. The spectral data are as follows: ^{1}H NMR (400 MHz, $CDCl_3$) δ: 1.37 (s, 12 H), 7.10 (dt, 1 H, J = 0.8, 8.7), 7.12 (d, 1 H, J = 0.7), 7.23 (dt, 1 H, J = 1.1, 7.6), 7.39 (dd, 1 H, J = 1.0, 8.3), 7.68 (dd, 1 H, J = 1.0, 8.1), 8.56 (br s, 1 H); ^{13}C NMR (100 MHz, $CDCl_3$) δ: 24.8, 84.1, 111.2, 113.8, 119.8, 121.6, 123.6, 128.3, 138.2; ^{11}B NMR (128.3 MHz, $CDCl_3$): δ 28.45 (BF_3•OEt_2 as external reference, δ 0.00); HRMS calcd for $C_{14}H_{18}BNO_2$[M^+], 243.1431, found 243.1423.

Safety and Waste Disposal Information

All hazardous materials should be handled and disposed of in accordance with "Prudent Practices in the Laboratory"; National Academy Press; Washington, DC, 1995.

3. Discussion

Aryl- and heteroarylboron derivatives are important class of compounds that have been applied to various fields of chemistry.[4] Traditional methods for their synthesis are based on the reactions of trialkylborates with aryl- and heteroarylmagnesium or lithium reagents.[5] Pd-catalyzed cross-coupling of aryl and heteroaryl halides with tetra(alkoxo)diborons[6] or di(alkoxo)boranes[7] is a milder variant where the preparation of magnesium and lithium reagents is avoided.

Alternatively, the catalytic C-H borylation of arenes and heteroarenes, first reported by Smith,[8] is highly attractive as a halide-free process for the synthesis of aryl- and heteroarylboron compounds.[9] Among the catalysts developed to date, the combination of $1/2[Ir(OMe)(COD)]_2$ and dtbpy described here exhibits high activity, which allows the formation of aryl- and heteroarylboronates in high yields at room temperature from an equimolar equivalent of bis(pinacolato)diboron (pin_2B_2, pin = $Me_4C_2O_2$) or pinacolborane (pinBH) and arenes or heteroarenes (Table 1).[9j-l]

The regiochemistry of arene borylation is primarily controlled by the steric effects of substituents. The reaction occurs at C-H bonds located *meta* or *para* to a substituent in preference to those located *ortho*. Thus, 1,2- and 1,4-disubstituted arenes bearing identical substituents yield arylboronates as single isomers. The borylation of 1,3-disubstituted arenes proceeds at the common *meta* position; therefore, regioisomerically pure products are obtained even for two distinct substituents on the arenes. In the case of five-membered heteroarenes, the electronegative heteroatom causes the C-H bonds at the α-positions to be active so that the borylation occurs at the α-positions. Thus, the regioselective monoborylation of 2-substituted or benzo-fused substrates can be possible. Although a mixture of 2-borylated and 2,5-diborylated products is formed from unsubstituted substrates, both products are selectively obtained by reactions with the appropriate ratio of substrate and reagent. On the other hand, six-membered heteroarenes such as pyridine shows significantly lower reactivity due to strong coordinating ability of the basic nitrogen for the catalyst. Exceptionally, 2,6-disubstituted pyridines undergo smooth borylation at the 4-position.

TABLE 1.

SYNTHESIS OF ARYL- AND HETEROARYLBORONATES[a]

Product	(%)[b] A[c]	(%)[b] B[d]	Product	(%)[b] A[c]	(%)[b] B[d]
3,4-Dichlorophenyl-Bpin	82 (8 h)	73 (8 h)	5-Methylthiophen-2-yl-Bpin	95 (2 h)	91 (2 h)
3,4-Dichlorophenyl-Bpin (Cl, Cl)	53 (24 h)	22 (24 h)	MeO_2C-furan-Bpin	80 (2 h)	95 (0.5 h)
3-Cl-5-I-phenyl-Bpin	82 (4 h)	67 (8 h)	Indol-2-yl-Bpin	88 (0.5 h)	99 (0.5 h)
3-F_3C-5-MeO-phenyl-Bpin	81 (8 h)	73 (24 h)	Thiophen-2-yl-Bpin	91[e] (1 h)	75[f] (0.5 h)
3-Cl-5-MeO_2C-phenyl-Bpin	80 (8 h)	70 (24 h)	Bpin-thiophene-Bpin	83[g] (0.5 h)	86[h] (2 h)
3-Br-5-NC-phenyl-Bpin	83 (2 h)	74 (2 h)	2-Cl-6-Me-pyridin-4-yl-Bpin	84 (4 h)	75 (2 h)

[a] Reactions were carried out at 25 °C in hexane (6 mL) by using substrate (1.0 mmol) and 1/2[Ir(OMe)(COD)]2-dtbpy catalyst. [b] GC yields based on substrates. [c] Method A: pin_2B_2 (0.50 mmol)/Ir cat. (0.015 mmol). [d] Method B: pinBH (1.1 mmol)/Ir cat (0.03 mmol). [e] 5.0 mmol of thiophene and 1.0 mmol of pinBH were used. The yield is based on pin_2B_2. [f] 5.0 mmol of thiophene was used. The yield is based on pinBH. [g] 1.0 mmol of pin_2B_2 was used. [h] 2.2 mmol of pinBH was used.

Functional group tolerance of the borylation is quite high. The reaction selectively occurs at the C-H bond for substrates possessing Cl, Br, I, CF_3, OMe, CO_2Me, and CN groups. The reaction occurs only at the aromatic C-H bonds even when the substrate has weaker benzylic C-H bonds.

Aryl- and heteroarylboronic acids and esters have been used for the synthesis of biaryls via the palladium-catalyzed cross-coupling reaction with aryl electrophiles.[10] Sequential reactions involving aromatic C-H borylation and cross-coupling with aryl electrophiles in the same flask provide an efficient and convenient route to unsymmetrical biaryls (eq. 1).

(1.3 mmol) — pin_2B_2 (0.65 mmol) or pinBH (1.4 mmol); cat Ir; hexane, 25 °C; pin = $Me_4C_2O_2$ — Br-C_6H_4-CO_2Me (1.0 mmol); cat $PdCl_2$(dppf); K_3PO_4; DMF/hexane; 60C, 2h — 91 - 96 %

1. Division of Molecular Chemistry, Graduate School of Engineering, Hokkaido University, Sapporo 060-8628, Japan.
2. Uson, R.; Oro, L. A.; Cabeza, J. A., *Inorg. Synth.* **1985**, *23*, 126.
3. (a) Nöth, H. Z. *Naturforsch. B: Anorg. Chem., Org. Chem.* **1984**, *39B*, 1463; (b) Ishiyama, T.; Murata, M.: Ahiko, T.-a; Miyaura, N. *Org. Synth.* **2000**, *77*, 176.
4. Reviews see: (a) Vaultier, M.; Carboni, B. in *Comprehensive Organometallic Chemistry II*; Abel, E. W., Stone, F. G. A., Wilkinson, G., Eds.; Pergamon Press: Oxford, 1995; Vol. 11, p 191; (b) Ishihara, K.; Yamamoto, H. *Eur. J. Org. Chem.* **1999**, 527; (c) Shinkai, S.; Ikeda, M.; Sugasaki, A.; Takeuchi, M. *Acc. Chem. Res.* **2001**, *34*, 494; (d) Entwistle, C.D.; Marder, T.B. *Angew. Chem., Int. Ed.* **2002**, *41*, 2927; (e) Soloway, A. H.; Tjarks, W.; Barnum, B. A.; Rong, F.-G.; Barth, R. F.; Codogni, I. M.; Wilson, J. G. *Chem. Rev.* **1998**, *98*, 1515; (f) Yang, W.; Gao, X.; Wang, B. *Med. Res. Rev.* **2003**, *23*, 346.
5. Nesmeyanov, A. N.; Sokolik, R. A. *Methods of Elemento-Organic Chemistry*; North-Holland: Amsterdam, 1967; Vol. 1.
6. (a) A review see: Ishiyama, T.; Miyaura, N. *J. Synth. Org. Chem., Jpn.* **1999**, *57*, 503; (b) A review see: Ishiyama, T.; Miyaura, N. *J. Organomet. Chem.* **2000**, 611, 392; (c) Isiyama, T.; Murata, M.; Miyaura, N. *J. Org. Chem.* **1995**, *60*, 7508; (d) Isiyama, T.; Itoh, Y.; Kitano, T.; Miyaura, N.; *Tetrahedron Lett.* **1997**, *38*, 3447; (e) Ishiyama, T.; Ishida, K.; Miyaura, N. *Tetrahedron* **2001**, 57, 9813.
7. Murata, M.; Oyama, T.; Watanabe, S.; Masuda, Y. *J. Org. Chem.* **2000**, *65*, 164.
8. Iverson, C. N.; Smith, M. R., III. *J. Am. Chem. Soc.* **1999**, *121*, 7696.

9. (a) A review see: Ishiyama, T.; Miyaura, N. *J. Organomet. Chem*. **2003**, *680*, 3; (b) Chen, H.; Hartwig, J. F. *Angew. Chem. Int. Ed.* **1999**, *38*, 3391; (c) Chen, H.; Schlecht, S.; Semple, T. C.; Hartwig, *J. F. Science* **2000**, *287*, 1995; (d) Cho, J.-Y.; Iverson, C. N.; Smith, M. R., III. *J. Am. Chem. Soc.* **2000**, *122,* 12868; (e) Tse, M. K.; Cho, J. Y.; Smith, M. R., III. *Org. Lett.* **2001**, *3*, 2831; (f) Cho, J.-Y.; Tse, M. K. Holmes, D.; Maleczka, R. E., Jr.; Smith, M. R., III. *Science* **2002**, *295*, 305; (g) Shimada, S.; Batsanov, A. S.; Howard, J. A. K.; Marader, T. B. *Angew. Chem., Int. Ed.* **2001**, *40*, 2168; (h) Isiyama, T.; Takagi, J.; Ishida, K,; Miyaura, N.; Anastasi, N. R.; Hartwig, J. F. *J. Am. Chem. Soc.* **2002**, *124*, 390; (i) Takagi, J.; Sato, K.; Hartwig, J. F.; Ishiyama, T.; Miyaura, N. *Tetrahedron Lett.* **2002**, *43*, 5649; (j) Isiyama, T.; Takagi, J.; Hartwig, J. F.; Miyaura, N. *Angew. Chem., Int. Ed.* **2002**, *41*, 3056; (k) Ishiyama, T.; Takagi, J.; Yonekawa, Y.; Hartwig, J. F.; Miyaura, N. *Adv. Synth. Catal.* **2003**, *345*, 1103; (l) Ishiyama, T.; Nobuta, Y.; Hartwig, J. F.; Miyaura, N. *Chem. Commun.* **2003**, 2924.
10. (a) Miyaura, N.; Suzuki, A. *Chem, Rev.* **1995**, *95*, 2457; (b) Miyaura, N. *Top. Curr. Chem.* **2002**, *219*, 11; (c) Suzuki, A.; Brown, H. C. *Organic Syntheses Via Boranes*; Aldrich Chemical Company, Inc.: Milwaukee, 2003; Vol. 3.

Appendix
Chemical Abstracts Nomenclature; (Registry Number)

Bis(η^4-1,5-cyclooctadiene)-di-μ-methoxydiiridium(l) ($[Ir(OMe)(COD)]_2$): Bis[(1,2,5,6-η)-1,5-cyclooctadiene]di-μ-methoxydiiridium; (12148-71-9)

4,4'-Di-*tert*-butyl-2,2'-bipyridine (dtbpy): 4,4'-Bis(1,1-dimethylethyl)-2,2'-bipyridine: (72914-19-3)

Bis(pinacolato)diboron: 4,4,4',4',5,5,5',5'-octamethyl-2,2'-Bi-1,3,2-dioxaborolane; (73183-34-3)

1-Chloro-3-iodobenzene; (625-99-0)

Pinacolborane: 4,4,5,5-Tetramethyl-1,3,2-dioxaborolane; (25015-63-8)

Indole: 1*H*-Indole; (120-72-9)

ASYMMETRIC REARRANGEMENT OF ALLYLIC TRICHLOROACETIMIDATES: PREPARATION OF (*S*)-2,2,2-TRICHLORO-*N*-(1-PROPYLALLYL)ACETAMIDE

(Acetamide, 2,2,2-trichloro-*N*-[(1*S*)-1-ethenylbutyl]-)

A. HO–(E)-hex-2-enyl–Me → (Cl_3CCN; DBU, CH_2Cl_2) → Cl_3C–C(=NH)–O–(E)-hex-2-enyl–Me

B. Cl_3C–C(=NH)–O–(E)-hex-2-enyl–Me → (2 mol% (*S*)-COP-Cl) → Cl_3C–C(=O)–NH–CH(propyl)–CH=CH$_2$

Submitted by Carolyn E. Anderson, Larry E. Overman,* and Mary P. Watson.
Checked by Matthew L. Maddess and Mark Lautens.

1. Procedure

Caution! Part A should be carried out in a well-ventilated hood to avoid exposure to trichloroacetonitrile vapors.

A. *Preparation of (E)-2,2,2-trichloroacetimidic acid hex-2-enyl ester.* A 500-mL, round-bottomed flask equipped with a stirring bar is flame-dried under a stream of nitrogen and allowed to cool to room temperature. The flask is then charged with *trans*-2-hexen-1-ol (Note 1) (3.3 mL, 27.8 mmol), 1,8-diazabicyclo[5.4.0]undec-7-ene (DBU) (Note 2) (0.84 mL, 5.6 mmol) and 170 mL of methylene chloride (Note 3). The solution is cooled to 4°C using an ice/water bath before trichloroacetonitrile (Note 2) (4.2 mL, 42 mmol) is added over five minutes by syringe. With time, the reaction solution is observed to change from clear to orange in color, and within 1 hour the starting materials are consumed (Note 4). The stir bar is removed with a magnetic rod and the solution is then concentrated under reduced pressure using a rotary evaporator to give a brown oil (Note 5). This oil is purified by flash chromatography through a plug of silica gel (Silicycle® 230-400 mesh, 6 cm tall, 5 cm diameter) using 2% ethyl acetate in hexanes (600 mL, 60 mL fractions) to yield 6.48 g (95%) of nearly pure (*E*)-2,2,2-trichloroacetimidic acid hex-2-enyl ester as a colorless oil (Notes 6 and 7).

B. *(S)-COP-Cl catalyzed rearrangement of (E)-2,2,2-trichloroacetimidic acid hex-2-enyl ester to (S)-2,2,2-trichloro-N-(1-propylallyl)acetamide.* A 150-mL, round-bottomed flask is fitted with a stirring bar and then charged with (*E*)-2,2,2-trichloroacetimidic acid hex-2-enyl ester (6.81 g, 27.8 mmol), di-μ-chlorobis[η^5-(*S*)-(p*R*)-2-(2'-(4'-isopropyl)oxazolinylcyclopentadienyl, 1-*C*, 3'-*N*))-(η^4-tetraphenylcyclobutadiene)cobalt]dipalladium [(*S*)-COP-Cl] (Note 8) (816 mg, 0.56 mmol) and 9.3 mL of methylene chloride (Note 3). The flask is sealed with a polyethylene cap, the cap is secured to the flask with parafilm, and the flask is placed in an oil bath preheated to 38°C +/- 2°C. After 24 h, the solution is cooled to room temperature, the stir bar is removed using a magnetic rod, and the solution is concentrated using a rotary evaporator to yield a brown oil. This oil is purified by flash chromatography through a column of silica gel (Silicycle® 230-400 mesh, 20 cm tall, 5 cm diameter) using 0.5% to 2% ethyl acetate:hexanes as eluent (3 L 0.5% ethyl acetate:hexanes, 1 L 1% ethyl acetate:hexanes, 1 L 2% ethyl acetate: hexanes). Evaporation of solvent provides 6.61 g (97% yield) of (*S*)-2,2,2-trichloro-*N*-(1-propylallyl)acetamide, 94% ee, as a pale yellow oil (Notes 9, 10, 11, and 12).

2. Notes

1. The checkers used *trans*-2-hexen-1-ol purchased from Aldrich Chemical Company, Inc. Although of sufficient purity for this series of transformations, the commercial reagent is contaminated with 3% of 1-hexanol.[1] The submitters used (*E*)-2-hexen-1-ol (>99% *E*) prepared from butanal by Horner-Wadsworth-Emmons reaction with trimethyl phosphonoacetate to form hex-2-enoic acid methyl ester, followed by reduction of this product with diisobutylaluminum hydride (–78°C to room temperature in THF).

2. 1,8-Diazabicyclo[5.4.0]undec-7-ene (DBU) and trichloroacetonitrile were purchased from Aldrich Chemical Company, Inc. These chemicals were used as received.

3. Methylene chloride was purified by passage through a solvent purification system. The submitters used a GlassContour alumina solvent purification columns and the checkers used a MBRAUN® solvent purification system.[2]

4. The reaction progress can be analyzed by thin layer chromatography (Silicycle®, plastic backed, 250 μm thickness). Using 10% ethyl acetate:hexanes as eluent, the product (*E*)-2,2,2-trichloroacetimidic acid hex-2-enyl ester has an R_f of 0.45, whereas the starting alcohol has an

R_f of 0.12. Both the imidate and starting alcohol can be visualized by potassium permanganate stain.

5. Upon concentrating the solution, black semi-solids sometimes form. These unwanted byproducts are insoluble in 2% ethyl acetate:hexanes.

6. The product, (*E*)-2,2,2-trichloroacetimidic acid hex-2-enyl ester, exhibits a spectrum that matches that which is reported in the literature.[3] ^{1}H NMR (300 MHz, $CDCl_3$) δ: 0.91 (t, J = 7.2 Hz, 3 H, CH_3), 1.43 (tq, J = 6.9, 7.2 Hz, 2 H, CH_2), 2.06 (tq, J = 6.9, 6.9 Hz, 2 H, CH_2), 4.74 (d, J = 6.3 Hz, 2 H, CH_2), 5.68 (dt, J = 15.6 Hz, 6.9 Hz, 1 H, C*H*), 5.86 (dt, J = 15.6, 6.3 Hz, 1 H, C*H*), 8.27 (broad s, 1 H, N*H*). ^{13}C NMR (75 MHz, $CDCl_3$) δ: 13.7, 22.1, 34.5, 70.0, 91.7, 123.3, 136.9, 162.6.

7. The product is contaminated with 2,2,2-trichloroacetimidic acid hexyl ester (3%) arising from the reaction of hexyl alcohol with trichloroacetonitrile.

8. (*S*)-COP-Cl was purchased from Aldrich Chemical Company, Inc. and was used as received. The specific rotation of this catalyst was measured to be: $[\alpha]_D^{23.8}$ = +1169 (c 0.25, $CHCl_3$).

9. The enantiomeric ratio was determined by HPLC analysis by comparison to a racemic sample:[3a] Hewlett-Packard HP Series 1100, Chiracel OD with guard column, 99:1 hexanes:isopropyl alcohol, 0.4 mL/min, 30 °C, R_t (major) = 15.3 min, R_t (minor) = 16.4 min.

10. ^{1}H NMR (400 MHz, $CDCl_3$) δ: 0.96 (t, J = 7.2 Hz, 3 H, CH_3), 1.35–1.46 (m, 2 H, CH_2), 1.55–1.71 (m, 2 H, CH_2), 4.39–4.47 (m, 1 H, C*H*), 5.17–5.27 (m, 2 H, CH_2), 5.81 (ddd, J = 16.4, 10.4, 5.6 Hz, 1 H, C*H*), 6.60 (broad s, 1 H, N*H*). ^{13}C NMR (100 MHz, $CDCl_3$) δ: 13.7, 18.9, 36.5, 53.4, 92.8, 115.9, 136.7, 161.2. IR (neat) cm^{-1}: 3424, 3322, 2961, 2935, 1714, 1520, 1249, 926, 821; HRMS (EI) calcd for $C_8H_{13}Cl_3NO$ (M+H) 244.0063, found 244.0063; R_f = 0.36 (10% ethyl acetate:hexanes).

11. The submitters reported, that when commercially available starting material is used, the product is contaminated with observable quantities of 2,2,2-trichloroacetimidic acid hexyl ester. The checkers, however, report that 2,2,2-trichloroacetimidic acid hexyl ester is not observed to the detection limits of ^{1}H or ^{13}C NMR.

12. On half-scale to that described, with prolonged drying under high vacuum the product was observed to solidify to a white crystalline solid (mp = 28 - 29°C). On the scale described above, the product remained an oil but could be rapidly induced to crystallize by addition of a seed crystal.

Safety and Waste Disposal Information

All hazardous materials should be handled and disposed of in accordance with "Prudent Practices in the Laboratory"; National Academy Press; Washington, DC, 1995.

3. Discussion

This procedure illustrates a general method for the preparation of enantioenriched chiral allylic trichloroacetamides from readily available prochiral (*E*)-allylic alcohols by catalytic asymmetric rearrangement of trichloroacetimidate intermediates.[4] The rearrangement tolerates a variety of alkyl and Lewis basic substituents at C3 of the starting allylic alcohol; however, substitution at C2 is not permitted (Table 1). [4,5] Although 5 mol % COP-Cl is convenient to use for small scale reactions, the catalyst loading

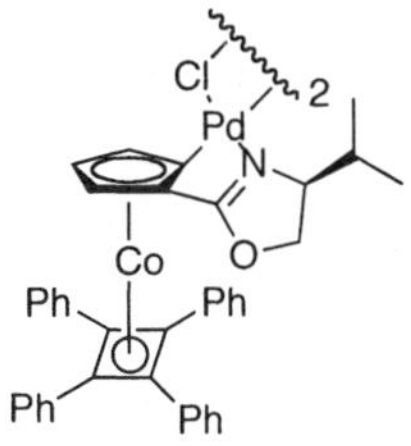

(*S*)-COP-Cl

can be decreased to as low as 1 mol % if the concentration is simultaneously increased (Entry 3). The reaction conditions exemplified in this example (2 mol % COP-Cl, CH_2Cl_2 (3M), 38°C, 24 h) were chosen to insure complete conversion of the allylic imidate intermediate within 24 hours. (*Z*)-Allylic trichloroacetimidates do not undergo COP-Cl catalyzed rearrangement at a practical rate. However, allylic trichloroacetamides of opposite absolute configuration can be prepared using (*R*)-COP-Cl, synthesized from (*R*)-valinol.

The COP-Cl catalyzed transformation of prochiral allylic alcohols to chiral allylic trichloroacetamides is technically simple, as allylic trichloroacetimidate intermediates require minimal purification. Additionally, no special precautions are required to protect the rearrangement reaction from light, air or traces of moisture. Moreover, the trichloroacetyl group of the product amide can be readily cleaved or this functional group can be converted to other functional arrays.[6]

Table 1. Enantioselective Synthesis of Allylic Trichloroacetimidates from (*E*)-Allylic Trichloroacetimidates.[a]

entry	R	amide yield (%)[b]	amide % ee[c]/conf
1	*n*-Pr	99	95/*S*
2	*i*-Bu	95	96/*S*
3[d]	*i*-Bu	92	98/*S*
4	CH_2CH_2Ph	93	93/*S*
5	$(CH_2)_3OAc$	97	92/*S*
6	$(CH_2)_2COMe$	98	95[e]/*S*
7	$CH_2OTBDMS$	98	96/*S*
8	$(CH_2)_3NBn(Boc)$	96	95/*S*

(a) Conditions: 5 mol % (*S*)-COP-Cl, CH_2Cl_2 (0.6 M), 38 °C, 18 h. (b) Duplicate experiments (±3%). (c) Determined by HPLC analysis of duplicate experiments (±2%). (d) 1 mol % (*S*)-COP-Cl, CH_2Cl_2 (1.2 M). (e) Determined by chiral GC analysis of duplicate experiments (±2%).

1. Hill, J. G.; Sharpless, K. B.; Exon, C. M.; Regenye, R. *Org. Syn., Coll. Vol. 7*, 461.
2. (a) Pangborn, A. B.; Giardello, M. A.; Grubbs, R. H.; Rosen, R. K.; Timmers, F. J. *Organometallics* **1996**, *15*, 1518–1520. (b) http://www.glasscontour.com/ or http://www.mbraunusa.com/.
3. (a) Overman, L. E. *J. Am. Chem. Soc.* **1976**, *98*, 2901–2910. (b) Bongini, A.; Cardillo, G.; Orena, M.; Sandri, S.; Tomasini, C. *J. Org. Chem.* **1986**, *51*, 4905–4910.
4. Anderson, C. E.; Overman, L. E. *J. Am. Chem. Soc.* **2003**, *125*, 12412–12413.
5. (a) Overman, L. E. *Acc. Chem. Res.* **1980**, *13*, 218–224. (b) Overman, L. E. *Angew. Chem., Int. Ed. Engl.* **1984**, *23*, 579–586.
6. For examples, see: (a) Cardillo, G.; Orena, M.; Sandri, S. *J. Chem. Soc., Chem. Commun.* **1983**, 1489-1490. (b) Nagashima, H.; Wakamatsu, H.; Itoh, K. *J. Chem. Soc., Chem. Commun.* **1984**, 652-653. (c) Atanassova, I. A.; Petrov, J. S.; Mollov, N. M. *Synthesis* **1987**, 734–736. (d) Yamamoto, N.; Isobe, M. *Chem. Lett.* **1994**, 2299–2302. (e) Urabe, D.;

Sugino, K.; Nishikawa, T.; Isobe, M. *Tetrahedron Lett.* **2004**, *45*, 9405-9407.

Appendix
Chemical Abstracts Nomenclature; (Registry Number)

trans-2-Hexen-1-ol: 2-Hexen-1-ol, (2*E*)-; (928-95-0)

1,8-Diazabicyclo[5.4.0]undec-7-ene (DBU): Pyrimido[1,2-a]azepine, 2,3,4,6,7,8,9,10-octahydro-; (6674-22-2)

Trichloroacetonitrile; (545-06-2)

(*E*)-2,2,2-trichloroacetimidic acid hex-2-enyl ester: Ethanimidic acid, 2,2,2-trichloro-, (2*E*)-2-hexenyl ester; (51479-70-0)

(*S*)-COP-Cl: Cobalt, bis[1,1',1",1'"-(η4-1,3-cyclobutadiene-1,2,3,4-tetrayl)tetrakis[benzene]](di-μ-chlorodipalladium)bis[μ-[(1-η:1,2,3,4,5-η)-2-[(4*S*)-4,5-dihydro-4-(1-methylethyl)-2-oxazolyl-κ*N*3]-2,4-cyclopentadien-1-yl]]di-; (581093-92-7)

PREPARATION OF (TRIPHENYLPHOSPHORANYLIDENE)-KETENE FROM (METHOXYCARBONYLMETHYLENE)-TRIPHENYLPHOSPHORANE

[Ethenone, (triphenylphosphoranylidene)-]

$$Ph_3P{=}CH{-}C(O)OCH_3 \xrightarrow[\text{toluene, 60°C, 24-30 hr}]{NaN[Si(CH_3)_3]_2} Ph_3\overset{+}{P}{-}\overset{-}{C}{=}C{=}O$$

Submitted by Rainer Schobert.[1]
Checked by Robert K. Boeckman, Jr. and Joseph E. Pero.

1. Procedure[2]

(Triphenylphosphoranylidene)ketene. A 3-L, three-necked, round-bottomed flask, equipped with a mechanical stirrer and a lateral gas inlet is purged with dry argon (Note 1) and charged with sodium amide (19.5 g, 0.5 mol) (Note 2), toluene dried by passage through alumina (1300 mL) (Note 3), and bis(trimethylsilyl)amine (81 g, 105 mL, 0.5 mol) (Note 4). The flask is fitted with a reflux condenser, then flushed once more with argon, and immersed in an oil bath. The mixture is heated at 70-80°C (bath temperature) until a clear and colorless solution of sodium hexamethyldisilazanide is obtained and the evolution of ammonia has ceased (2-4 hr) (Notes 5, 6). The reaction mixture is cooled to room temperature and the condenser is removed briefly, with protection from air by a stream of argon, to allow the addition of (methoxycarbonylmethylene)triphenylphosphorane (167 g, 0.5 mol) in several large portions (Note 7). The mixture is then heated again to ca. 60-70°C (bath temperature) and stirring is continued for a further 24-30 hr by which time the solution should take on a bright yellow hue (Note 8). While still hot and blanketed with a stream of argon, the reaction mixture is rapidly suction filtered using a 24/40 standard taper vacuum filtration adapter and a 350-mL fritted funnel (course porosity) open to the atmosphere through a 1 inch thick pad of Celite™ into a 2-L 24/40 single-necked round-bottomed flask (Notes 9 and 10).

Once filtration is complete, the receiving flask is flushed with argon and attached to a rotary evaporator (Note 11). The colorless filtrate is evaporated to dryness and the residue recrystallized from dry toluene (~1g /

5mL) by dissolution in hot toluene followed by cooling to -20°C in a freezer overnight. While the mixture was cold, the resulting solids are collected by suction filtration using a 350-mL coarse fritted funnel and washed on the filter with three 100 mL portions of cold (0°C) toluene affording the first crop of phosphorane (Note 3). The mother liquor is again concentrated to dryness under vacuum, redissolved in ~100 mL of hot dry toluene, and again cooled to -20°C in a freezer overnight. The resulting solids are isolated and washed as above. The first and second crops of phosphorane are then combined and dried under high vacuum (ca 0.01 Torr) to a constant weight to afford 106.8-114.9 g (74-76%) (Note 12) of the phosphorane product as a very pale yellow, flaky powder having mp 173°C (Note 13).

2. Notes

1. An atmosphere of dry argon is maintained throughout the entire process. While neither the starting nor the product ylide are particularly air- or moisture-sensitive in the solid state, hot solutions of the latter are sensitive.

2. Sodium amide should be of 95% purity. Powdery batches were found to react more readily than suspensions in mineral oil or pellets. Sodium amide can be purchased from Aldrich Chemical Company, Inc. The checkers employed sodium amide purchased from Acros Organics CAUTION: Sodium amide is highly flammable and corrosive and can cause severe burns.

3. The checkers employed toluene dried by passage through a column containing anhydrous alumina under nitrogen pressure. The submitters dried toluene by distillation from sodium metal. Attempts by the checkers to employ toluene dried by distillation from CaH_2 led to inferior results during attempts to form sodium hexamethyldisilazanide (NaHMDS).

4. Bis(trimethylsilyl)amine (hexamethyldisilazane, HMDS), 99+%, was purchased from Aldrich Chemical Company, Inc. CAUTION: bis(trimethylsilyl)amine is harmful and corrosive.

5. Purging the set-up with argon every now and then cuts down reaction times considerably.

6. Commercially available sodium hexamethyldisilazanide (e.g. Aldrich Chemical Company) can be used instead. However, it is costly and the reaction time and yields of product crucially depend on its age and quality.

7. (Methoxycarbonylmethylene)triphenylphosphorane can be bought from Aldrich Chemical Company, Inc. The checkers employed Aldrich

material as received. The phosphorane can also be readily prepared at low cost in two steps from triphenylphosphine and methyl bromoacetate according to a literature procedure.[3]

8. Monitoring reaction progress is possible by IR spectroscopy of aliquots of the reaction mixture measured as films on KBr plates. A strong product band at 2090 cm^{-1} of constant intensity and the disappearance of a starting ylide band at 1616 cm^{-1} are indicative of a complete reaction.

9. The submitters employed a funnel with a sintered frit (fine porosity) where the frit is covered with a 2 cm layer of basic alumina (III) and 1 cm of celite atop. Through side-arm stopcocks, the filter funnel is connected to an argon line and the receiving flask to a source of vacuum (water-jet or membrane pump). The whole set-up is positioned upright or slightly slanted and submitted to several cycles of evacuating and flushing with inert gas prior to use. During filtration occasional removal or resuspending of sedimented sodium methoxide from the surface of the celite/alumina plug by scraping with a long spatula may be necessary in order to maintain a continuous flow of filtrate.

10. The submitters filtered the reaction mixture by pouring through a tapered adaptor into a 2-L Schlenk-type filter funnel which is joined to the only neck of a round-bottomed 2-L receiving flask (Note 9). Filtration is then carried out by application of a slight overpressure from an argon line to the stoppered funnel and by gentle suction at the receiving flask (the stopcock to the vacuum source should be opened at regular intervals, but only briefly at each time). During filtration which typically takes 30 to 60 min, the temperature of the reaction mixture in the filter funnel should be kept between 30°C and 50°C by means of a heating basket, mantle or tape, or at least insulating foil to avoid premature product crystallization.

11. The submitters concentrated the clear filtrate to 500 mL, flushed once more with argon and then placed in a refrigerator for 24 hr. The faintly yellow precipitate (ca. 70-80 g) is collected on a second Schlenk-type filter funnel, washed with dry diethyl ether (300 mL) and dried on an oil pump at ca. 0.01 Torr. A second, darker crop of product (ca. 20-35 g) is obtained by concentrating the toluene mother liquor to ca. 250 mL, layering it with diethyl ether (50 mL) and cooling the mixture to 4°C for 24 hr. The resulting precipitate is collected, washed and dried like the first batch to which it is by then, identical in color and all analytical data.

12. The submitters obtained 100-115 g (66-75%) of pure phosphorane in two crops with identical mp and spectroscopic properties using the above described direct crystallization procedure (Note 11).

13. The product is pure by usual microanalytical standards; IR (KBr,

ν): 2090 (s), 1625 (m), 1436 (m), 1110 (m) cm^{-1}; ^{1}H NMR ($CDCl_3$, 400 MHz) δ: 7.44-7.54 (m, 6H), 7.55-7.63 (m, 3H), 7.64-7.77 (m, 6H); ^{13}C NMR ($CDCl_3$, 100 MHz) δ: -10.5 (d, $^{1}J_{PC}$ 185.4 Hz, C^{α}), 128.8 (d, $^{3}J_{PC}$ 12.9 Hz, C-*ortho*), 129.6 (d, $^{1}J_{PC}$ 98.5 Hz, C-*ipso*), 132.2 (s, C-*para*), 132.3 (s, C-*meta*), 145.6 (d, $^{2}J_{PC}$ 43.0 Hz, C^{β}); ^{31}P NMR ($CDCl_3$, 162 MHz) δ: 6.0. Analysis performed on an unrecrystallized sample of the bulk material: Anal. Calcd for $C_{20}H_{15}OP$: C, 79.46; H, 5.00; P, 10.25. Found. C, 78.66; H, 5.30. Recrystallization of this sample from toluene and reanalysis gave: Anal. Calcd for $C_{20}H_{15}OP$: C, 79.46; H, 5.00; P, 10.25. Found. C, 79.60; H, 5.06. The only detectable impurities by ^{1}H NMR are the starting ester ylide and toluene (δ 3.60, s and 2.36, s). Based on this intergration, the purity of product is calculated to be ~99.97%. Prior analytical results have been reported.[4-6]

Waste Disposal Information

All hazardous materials should be handled and disposed of in accordance with "Prudent Practices in the Laboratory"; National Academy Press; Washington, DC, 1995.

3. Discussion

Alternative methods for the preparation of Ph_3PCCO: Strong bases other than sodium hexamethyldisilazanide (NaHMDS) have been employed for the elimination of methanol from (methoxycarbonyl-methylene)triphenylphosphorane. With crystalline sodium amide, drastic reaction conditions (refluxing in benzene or toluene for several days) are required.[7] However, mixtures of sodium amide and catalytic quantities of HMDS react as readily as stoichiometric amounts of preformed NaHMDS.[8] Alkyllithium bases (e.g. *n*BuLi, PhLi) give low yields of Ph_3PCCO contaminated with various byproducts stemming from attack of the base on the phenyl rings.[9] The title ylide has also been prepared by pyrolysis of the disilylated ester ylide obtained from reaction of [1,1-bis(trimethylsilyl)methylene]triphenylphosphorane with carbon dioxide,[10] and by pyrolysis of the betaine obtained from the reaction of carbodiphosphorane, $Ph_3P{=}C{=}PPh_3$, with carbon dioxide.[6] However, neither protocol is suitable for large scale preparation.

Properties and reactions of Ph_3PCCO:[11] It is remarkably stable and solid samples can be stored for months at room temperature under an atmosphere of argon and handled even under ambient conditions without notable decomposition. This fact is due to its electronic structure featuring an orthogonal pair of filled π-systems ($\pi^4 \perp \pi^4$). Hence Ph_3PCCO shows little propensity to dimerize like the strongly electrophilic ketenes. It does not even enter readily into Wittig alkenation reactions which are so typical of common phosphorus ylides. It represents a nucleophilic-only C_2-building block reacting with a wide range of electrophiles in a variety of ways and undergoing cycloadditions both at the polar $P–C^{\alpha}$ and at the $C^{\alpha}=C^{\beta}$ bond. For example, acidic compounds, RXH, such as alcohols, thiols, amines, and CH-active 1,3-dicarbonyls add across the C=C bond of Ph_3PCCO to give the corresponding Wittig-active acyl ylides, $Ph_3P=CH–C(O)XR$. Multicomponent and domino reactions between the cumulated ylide, an

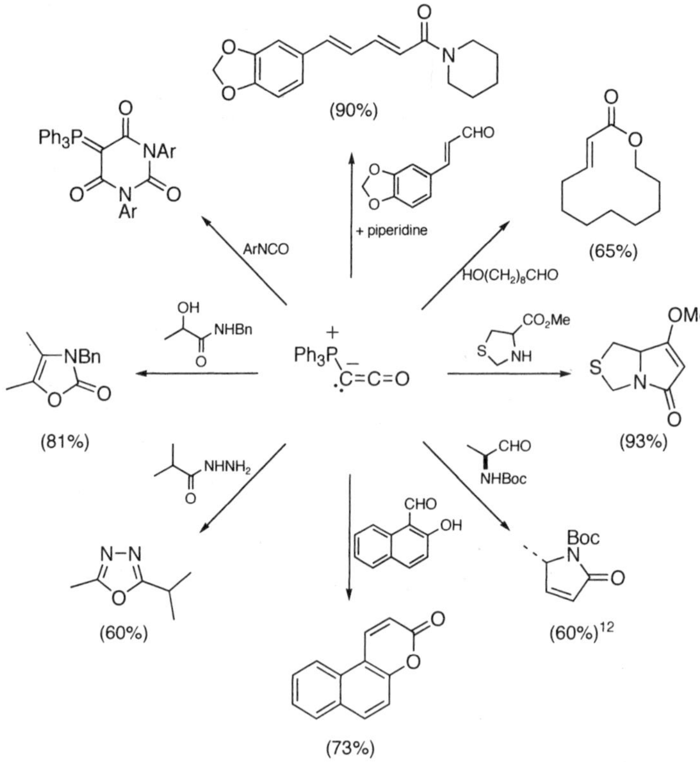

acidic compound, and a ketone or an aldehyde lead to various α,β-unsaturated carbonyl compounds. Intramolecular variants employing difunctional carbonyl derivatives, including esters and amides, bearing such acidic functionalities furnish (hetero)cyclic products with ring sizes ranging from five to twenty. Grignard compounds, RMgX, react with Ph_3PCCO to yield acyl ylides $Ph_3P{=}CHC(O)R$ upon hydrolytic work-up. Both [2+2]- and [2+4]-reactions are known of Ph_3PCCO, e.g. with ketenes to form cyclobutan-1,3-diones, and with iso(thio)cyanates to furnish six-membered pyrimidinetriones, or hexahydro-1,3-dithiazines, respectively.

1. Universität Bayreuth, Organisch-chemisches Laboratorium, Universitätsstrasse 30, D-95440 Bayreuth (Germany). E-mail: Rainer.Schobert@uni-bayreuth.de, Fax: +49 (0)921 552672.
2. Bestmann, H. J.; Sandmeier, D. *Angew. Chem.* **1975**, *87*, 630; *Angew. Chem., Int. Ed. Engl.* **1975**, *14*, 634.
3. Isler, O.; Gutmann, H.; Montavon, M.; Ruegg, R.; Ryser, G.; Zeller, P. *Helv.Chim. Acta* **1957**, *40*, 1242-1249.
4. Daly, J. J.; Wheatley, P. J. *J. Chem. Soc. A* **1966**, 1703-1706.
5. Matthews, C. N.; Birum, G. H. *Chem. Ind. (London)* **1968**, 653.
6. (a) Matthews, C. N.; Birum, G. H. *Tetrahedron Lett.* **1966**, 5707-5710; (b) Matthews, C. N.; Birum, G. H. *Acc. Chem. Res.* **1969**, *2*, 373-379.
7. Bestmann, H. J.; Schmidt, M.; Schobert, R. *Synthesis* **1988**, 49-53.
8. Haas, W. *PhD Thesis*, University of Erlangen-Nürnberg, Germany **1992**, 20;
9. (a) Bestmann, H. J.; Besold, R.; Sandmeier, D. *Tetrahedron Lett.* **1975**, 2293-2294; (b) Buckle, J.; Harrison, P. G. *J. Organomet. Chem.* **1974**, *77*, C22-C24.
10. Bestmann, H. J.; Dostalek, R.; Zimmermann, R. *Chem. Ber.* **1992**, *125*, 2081-2084.
11. (a) Bestmann, H. J. *Angew. Chem.* **1977**, *89*, 361-376; *Angew. Chem., Int. Ed. Engl.* **1977**, *16*, 349-364; (b) Schobert, R.; Gordon, G. J. *Science of Synthesis: Houben-Weyl Methods of Molecular Transformations*; Thieme: Stuttgart, **2004**, Vol. 27, p. 1042.
12. Andrus, M. B.; Li, W.; Keyes, R. F. *J. Org. Chem.* **1997**, *62*, 5542-5549.

Appendix
Chemical Abstracts Nomenclature (Collective Index Number); (Registry Number)

(Triphenylphosphoranylidene)ketene: Phosphonium, triphenyl-, oxoethenylide (9); (73818-55-0) *and* Ethenone, (triphenylphosphoranylidene)- (9); (15596-07-3)

(Methoxycarbonylmethylene)triphenylphosphorane: Phosphonium, triphenyl-, 2-methoxy-2-oxoethylide (9); (21204-67-1)

Hexamethyldisilazane: Silanamine, 1,1,1-trimethyl-*N*-(trimethylsilyl)- (9); (999-97-3)

Sodium Amide (9); (7782-92-5)

PREPARATION OF [1-(METHOXYMETHYLCARBAMOYL)ETHYL]PHOSPHONIC ACID BIS(2,2,2-TRIFLUOROETHYL) ESTER: A USEFUL INTERMEDIATE IN THE SYNTHESIS OF *Z*-UNSATURATED *N*-METHOXY-*N*-METHYLAMIDES

A. 1 → 2 (TFE, TEA, THF, 0 °C to r.t., 2.5 hr)

B. 3 → 4 (Triphosgene, Pyridine, DCM, -78 °C to r.t., 16.5 hr)

C. 2 + 4 → 5 (nBuLi, HMDS, THF, -78 °C, 2.5 hr)

Submitted by Amos B. Smith, III, Jason J. Beiger, Akin H. Davulcu, and Jason M. Cox.[1]

Checked by Mark Lautens and Catherine Taillier.

1. Procedure

A. *Ethylphosphonic acid bis(2,2,2-trifluoroethyl) ester* (**2**).[2] A flame-dried 2-L, three-necked, round-bottomed flask equipped with a Teflon-coated magnetic stir bar, thermometer, rubber septum and argon inlet (Note 1) is charged with 24.1 mL (336 mmol, 2.01 eq.) of 2,2,2-trifluoroethanol (Notes 2,3) and 500 mL of anhydrous THF (Note 4) under an argon atmosphere, and cooled to an internal temperature of 0°C with an ice-NaCl bath for 10 min. The above flask is then charged dropwise with 51.3 mL (369 mmol, 2.20 eq.) of triethylamine (Note 5) over 10 min., and stirred at 0°C for 15 min. A separate flame-dried 250-mL, one-necked, round-bottomed flask equipped with a rubber septum and argon inlet is charged with 17.9 mL (167 mmol, 1.00 eq.) of ethylphosphonic dichloride 1 (Note 2) and 65 mL of THF under an argon atmosphere. The mixture is stirred, and transferred dropwise *via* cannula (Notes 3 and 6) into the trifluoroethanol/triethylamine solution over 15 min while maintaining the internal temperature below 12°C. The 250-mL flask is then rinsed with THF (2 x 10 mL), and the THF washes transferred by cannula to the larger flask over a five-minute period. The resulting white slurry is warmed to room temperature over 15 min., stirred at room temperature for 2.25 h, filtered through a sand-covered frit (Note 7), and concentrated by rotary evaporation (35°C/20 mmHg) to afford 43 g of a cloudy, nearly colorless liquid. The latter is transferred to a 100-mL, one-necked, round-bottomed flask equipped with a Teflon-coated magnetic stir bar and a short path distillation head fitted with a distribution adapter and three 50-mL pear-shaped receiving flasks. After collecting a forerun (ca. 1 mL), the product is distilled under an aspirator vacuum (bp = 92–95°C/25 mmHg) to yield 40.85 g (89%) of ethylphosphonic acid bis(2,2,2-trifluoroethyl) ester (**2**) as a clear colorless liquid (Note 8).

B. *N-Methoxy-N-methylcarbamoyl chloride* (**4**).[3] A flame-dried 500-mL, three-necked, round-bottomed flask equipped with a Teflon-coated magnetic stir bar, thermometer, rubber septum and argon inlet is charged with 10.1 g (34.1 mmol, 0.400 eq.) of triphosgene (Note 2) and 57 mL of anhydrous dichloromethane (Note 4) under an argon atmosphere, and cooled to an internal temperature of –78°C with a dry ice-acetone bath for 20 min. The suspension is charged with 8.33 g (85.4 mmol, 1.00 eq.) of *N,O*-dimethylhydroxylamine hydrochloride **3** (Note 2) in three portions by briefly removing the septum and using a powder funnel with 5 min. elapsing between each portion while maintaining the internal temperature below –70°C. The suspension is recooled to –78°C and stirred for 30 min. A separate flame-dried, 100-mL round-bottomed flask equipped with a rubber septum and argon inlet is charged with 13.7 mL (171 mmol, 2.00 eq.) of pyridine (Note 5) and 28 mL of dichloromethane under an argon atmosphere. The mixture is stirred and transferred dropwise *via* cannula to the above hydroxylamine/triphosgene suspension over 1.5 h while maintaining the internal temperature below –72°C (Note 6). The 100-mL flask is rinsed with dichloromethane (2 x 5 mL), and the dichloromethane washes transferred by cannula to the larger flask over 5 min. The resulting yellow-orange slurry is slowly warmed to room temperature over 4 h, stirred at room temperature for 12.5 h, quenched with 100 mL of distilled water, and transferred to a 500-mL separatory funnel. The layers are separated, and the aqueous layer extracted with dichloromethane (3 x 50 mL). The combined organic phases are sequentially washed (Note 5) with aqueous 0.5*M* HCl (2 x 100 mL), saturated aqueous $NaHCO_3$ (1 x 75 mL), and brine (1 x 75 mL), dried (Na_2SO_4), filtered, and concentrated by rotary evaporation (35°C/20 mmHg) to afford 11.3 g of a yellow liquid. The yellow liquid is transferred to a 50-mL, one-necked, round-bottomed flask equipped with a Teflon-coated magnetic stir bar and a short path distillation head fitted with a distribution adapter and three 25-mL pear-shaped receiving flasks. After collecting a forerun (*ca.* 0.5 mL), the product is distilled under vacuum aspirator (bp = 67–69°C/25 mmHg) to afford 8.87-9.18 g (84-87%) of *N*-methoxy-*N*-methylcarbamoyl chloride **4** as a clear colorless liquid (Note 9).

C. *[1-(Methoxymethylcarbamoyl)ethyl]phosphonic acid bis(2,2,2-trifluoroethyl) ester* (**5**).[4] A flame-dried, 1-L, three-necked, round-bottomed flask equipped with a Teflon-coated magnetic stir bar, thermometer, rubber septum and argon inlet is charged with 51.0 mL (128 mmol, 2.41 eq.) of a 2.5*M* solution of *n*-butyllithium in hexanes (Note 2) and 40 mL of THF under an argon atmosphere, and cooled to an internal temperature of –20 ± 3°C with a dry ice-isopropanol-water bath for 20 min. In a separate flame-dried, 100-mL, one-necked, round-bottomed flask equipped with a rubber septum and argon inlet is charged with 29.2 mL (139 mmol, 2.64 eq.) of 1,1,1,3,3,3-hexamethyldisilazane (HMDS) (Note 2) and 40 mL of THF under an argon atmosphere. The mixture is stirred and transferred dropwise *via* cannula to the above *n*-butyllithium solution over 20 min while maintaining the internal temperature below –15°C (Note 6). The 100-mL flask is then rinsed with THF (2 x 10 mL), and the THF washes cannulated to the larger flask over 5 min. The resulting clear solution is stirred at –20 ± 3°C for 20 min., and cooled to an internal temperature of –75°C with a dry ice-acetone bath. A separate flame-dried, 100-mL, one-necked, round-bottomed flask equipped with a rubber septum and argon inlet is then charged with 14.5 g (52.9 mmol, 1.00 eq.) of ethylphosphonic acid bis(2,2,2-trifluoroethyl) ester **2**,[2] 9.00 g (72.9 mmol, 1.38 eq.) of *N*-methoxy-*N*-methylcarbamoyl chloride **4**,[3] and 50 mL of THF, and transferred dropwise *via* cannula to the above lithium HMDS solution over 30 min. while maintaining the internal temperature below –68°C (Note 6). The smaller flask is rinsed with THF (2 x 10 mL), and the THF washes transferred by cannula to the larger flask over 5 min. The resulting pale yellow solution is recooled to –75°C and stirred for 2.5 h (Note 10), then slowly acidified over 5 min. with 130 mL of a 1.0 *M* solution of HCl, and gradually warmed to an internal temperature of 0°C over 30 min. The solution is diluted with 100 mL of distilled water, and transferred to a 1-L separatory funnel. The flask is rinsed with diethyl ether (3 x 50 mL), transferred to the funnel, shaken, the layers separated, and the aqueous layer extracted with dichloromethane (4 x 100 mL) (Note 11). The combined organic phases are dried ($MgSO_4$), filtered, and concentrated by rotary evaporation (35°C, 20 mmHg) to afford 19 g of a yellow oil (Note 12). The yellow oil is loaded onto a 80-mm diameter column, wet-packed (4:1 hexanes:ethyl acetate) with 450 grams (25 cm) of silica gel (Note 13), and sequentially eluted with a gradient of hexanes and ethyl acetate (2 L of 2:1, 2 L of 1:1, 1 L of 1:2). The desired product is collected in fractions of 75-mL volume, concentrated by rotary evaporation (35°C/20 mmHg), and dried under vacuum (25°C/0.01 mmHg) until a constant mass is obtained. The above described procedure affords 15.6 - 16.8 g (82–88%) of [1-(methoxymethylcarbamoyl)ethyl]phosphonic acid bis(2,2,2-trifluoroethyl) ester **5** as a clear pale yellow oil (Note 14).

2. Notes

1. All flasks were flame-dried and maintained under an argon atmosphere during the course of the reactions. The argon was dried and purified by passing through Drierite® and then an Oxiclear™ disposable gas purifier, which may be purchased from Aldrich Chemical Co., Inc. A gas manifold was then used to distribute argon to each of the attached flasks. An argon inlet was affixed to a flask by inserting its needle through that flask's rubber septum while maintaining slightly positive pressure.

2. 2,2,2-Trifluoroethanol (≥99%), ethylphosphonic dichloride (98%), triphosgene (98%), *N,O*-dimethylhydroxylamine hydrochloride (98%), 1,1,1,3,3,3-hexamethyldisilazane (99.9%) and the 2.5M solution of *n*-butyllithium in hexanes were purchased from Aldrich Chemical Co., Inc. and used as received.

3. Liquids were added to flasks from plastic disposable syringes through stainless steel needles. Liquids were transferred between flasks *via* stainless steel cannulas. All needles and cannulas were oven-dried for at least one hour and cooled to room temperature in a dessicator prior to use.

4. HPLC grade dichloromethane (99.9%) and tetrahydrofuran (THF) (99.9%) were purchased from Fisher Scientific. Dichloromethane and THF were dried by purging with argon over activated molecular sieves and were stored under argon at room temperature. The checkers used THF distilled from Na/benzophenone ketyl and dichloromethane purified with a MBRAUN® Solvent Purification System.

5. Triethylamine (99%), pyridine (99.9%), dichloromethane, hexanes, ethyl acetate, sodium chloride, sodium bicarbonate, anhydrous magnesium sulfate, anhydrous sodium sulfate, and concentrated hydrochloric acid were purchased from Fisher Scientific and used as received. Triethylamine and pyridine were used by the checkers after distillation over KOH.

6. Alternatively, a flame-dried, 100-mL additional funnel may be used to regulate the rate at which the reaction flask is charged with solution.

7. Alternatively, filtration of the resulting white slurry through a short plug of Celite could be performed to remove the ammonium salts.

8. Ethylphosphonic acid bis(2,2,2-trifluoroethyl) ester (**2**) displays the following spectroscopic properties: IR (film): 2974 (weak), 1463, 1419, 1287, 1255, 1174, 1109, 1079, 1036, 964, 868, 844 cm^{-1}; ^{1}H NMR (400 MHz, $CDCl_3$) δ: 1.22 (dt, J_{PH} = 21.5 and J_{HH} = 7.7 Hz, 3 H), 1.93 (dq, J_{PH} = 18.7 and J_{HH} = 7.7 Hz, 2 H), 4.30-4.47 (m, 4 H); ^{13}C NMR (100 MHz, $CDCl_3$) δ: 5.9 (d, J_{CP} = 7.7 Hz), 18.9 (d, J_{CP} = 143.4 Hz), 61.8 (qd, J_{CF} = 37.6 and J_{CP} = 6.1 Hz), 122.6 (qd, J_{CF} = 277.6 and J_{CP} = 7.7 Hz); The checkers have not been able to get an elemental analysis due to volatility of compound **2**. However, the submitters reported the following data for **2**: Anal. Calcd for $C_6H_9F_6O_3P$: C, 26.29; H, 3.31. Found: C, 26.20; H, 3.06.

9. The following characterization data were obtained for *N*-methoxy-*N*-methylcarbamoyl chloride (**4**): IR (film): 2982, 2941, 1732 (broad), 1460, 1443, 1406, 1352, 1182, 1082, 995, 868, 669, 653 cm^{-1}; ^{1}H NMR (500 MHz, $CDCl_3$): δ 3.33 (s, 3H), 3.78 (s, 3H). The checkers have not been able to get an elemental analysis due to volatility of compound **4**. However, the submitters reported the following data for **4**: Anal. Calcd for $C_3H_6ClNO_2$: C, 29.17; H, 4.90. Found: C, 29.02; H, 4.89.

10. The reaction was monitored by thin layer chromatography (TLC) with 0.25-mm E. Merck pre-coated silica gel plates. The plates were eluted with a 1:1 mixture of hexanes and ethyl acetate. The R_f values of **2**, **4**, and **5**, are 0.55, 0.70, and 0.21, respectively. A potassium permanganate-based stain that was prepared by dissolving 3 g of $KMnO_4$, 20 g of K_2CO_3, and 5 mL of 5% aqueous NaOH in 300 mL of distilled water was suitable for visualizing the product on TLC.

11. If an emulsion occurs during the dichloromethane extractions, it may be broken with the addition of approximately 15 mL of brine. Checkers found that filtration of the biphasic mixture through a short plug of Celite was necessary to break the emulsion.

12. The crude mixture may be distilled under vacuum through a short path apparatus (bp = 105–107°C/0.01 mmHg) to yield phosphonate **5** as a

slightly pale yellow oil. Purification by this method, however, resulted in a significantly lower yield (40–50%) due to thermal decomposition of **5**.

13. Silica gel may be purchased from Silicycle Chemical Division. The silica gel used by the submitters had the following specifications: pH: 6.5–7.0, particle size: 40–63 μm (230–400 mesh), spec. surface area: 500 m^2/g, pore diameter: 60 Å.

14. Phosphonate **5** exhibits the following spectroscopic properties: IR (film): 2976, 2949, 1651, 1455, 1418, 1393, 1259, 1173; 1072, 987, 962, 845 cm^{-1}; ^1H NMR (400 MHz, $CDCl_3$) δ: 1.50 (dd, J = 19.8 and J = 7.3 Hz, 3 H), 3.28 (s, 3 H), 3.77 (m, 1 H), 3.78 (s, 3 H), 4.35-4.47 (m, 2 H), 4.43-4.57 (m, 2 H); ^{13}C NMR (100 MHz, $CDCl_3$) δ: 12.4 (d, J_{CP} = 7.7 Hz), 32.0, 35.7 (d, J_{CP} = 142.6 Hz), 61.7, 61.8 (qd, J_{CP} = 38.3 and J_{CP} = 6.1 Hz), 63.1 (qd, J_{CP} = 37.6 and J_{CP} = 5.0 Hz), 122.6 (qdd, J_{CF} = 277.6, J_{CP} = 22.6 and J_{CP} = 8.8 Hz), 169.3; MS-ESI *m/z* (relative intensity): 284 ($M+Na^+$, 100%), 362 ($M+H^+$, 81%), 319 (27%), 301 (22%), 273 (25%), 245 (8%); High resolution mass spectrum (ES+) *m/z* 384.0393 [$(M+Na)^+$; calcd for $C_9H_{14}F_6NNaO_5P$: 384.0411]. Anal. Calcd for $C_9H_{14}F_6NO_5P$: C, 29.93; H, 3.91; N, 3.88. Found: C, 29.97; H, 3.93; N, 3.92.

Waste Disposal Information

All hazardous materials should be handled and disposed of in accordance with "Prudent Practices in the Laboratory"; National Academy Press; Washington, DC, 1995.

3. Discussion

The methods described herein illustrate a practical and convenient three-step synthesis of bis(2,2,2-trifluoroethyl)phosphonates in good overall yield (>66%). This class of phosphonates, bearing an *N*-methoxy-*N*-methylamide functionality, is synthetically useful for the construction of *Z*-unsaturated *N*-methoxy-*N*-methylamides.[4] It is well established that the Weinreb amide functionality may then be converted to aldehydes or ketones in good yields.[5]

The formation of bis(2,2,2-trifluoroethyl)phosphonate **5** was previously reported by Deslongchamps et al.[4] following procedures developed by Tius and Busch-Petersen that employed KF/alumina to construct α-heterosubstituted Weinreb amides.[6] They discovered that tris(2,2,2-trifluoroethyl) phosphite reacts with bromides bearing the *N*-

methoxy-*N*-methylamide moiety (**6**, **7**) to furnish the respective phosphonates **8** and **5** in modest yield (eq. 1).

6 R = H
7 R = Me

$(CF_3CH_2O)_3P$
KF/alumina
MeCN, reflux
33%

8 R = H
5 R = Me

(1)

The procedure described herein proceeds in higher yield and circumvents the difficulty of handling and preparing the KF/alumina reagent. For example, although reactivity of the halide increases as the ratio of fluoride salt versus alumina is increased, the reagents became very hygroscopic and difficult to handle.[7] Additionally, Ando *et al.* discovered that a trace amount of water was essential for promoting the reaction, and thorough drying of the KF/alumina reagent led to a significant reduction in reactivity.[6,8] On the other hand, too much residual water was found to hydrolyze the chlorine analog of Weinreb amide **6**.[6] Tius and Busch-Petersen also reported that continuous sonication of the reaction mixture was necessary to enhance the reactivity of the KF/alumina reagent.[6,9]

8 R = H
5 R = Me

Aldehydes **9–12**
18-crown-6,
KH, THF, –78 °C
81–89%

13–17

(2)

Deslongchamps *et al.* subsequently demonstrated the utility of phosphonates **5** and **8** in the stereoselective formation of *Z*-unsaturated alkenes using the modified Still conditions.[4,10] Bis(2,2,2-trifluoroethyl)phosphonates **8** and **5** react with aldehydes **9–12** to furnish the respective unsaturated *N*-methoxy-*N*-methylamides **13–17** in 81–89% yield, with only the *Z* isomer detectable by ^{1}H-NMR (eq. 2).[4]

Table 1 illustrates the several aldehydes (**9–12**) that Deslongchamps *et al.* employed to demonstrate the versatility of bis(2,2,2-trifluoroethyl)phosphonates **5** and **8** in the stereoselective synthesis of *Z*-unsaturated *N*-methoxy-*N*-methylamides **13–17**.[4]

Table 1. Synthesis of *Z*-Unsaturated-*N*-methoxy-*N*-methylamides.

Entry	Aldehyde	Phosphonate	Product	Yield
1	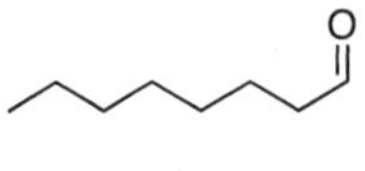 9	5	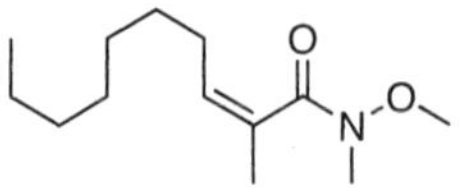 13	83
2	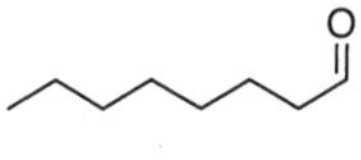 9	8	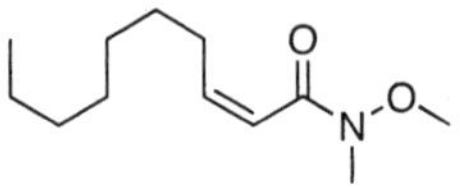 14	89
3	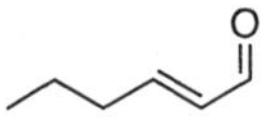 10	5	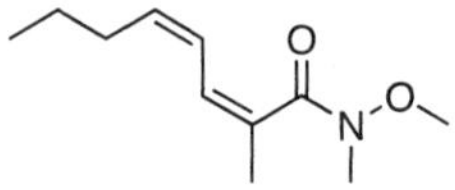 15	83
4	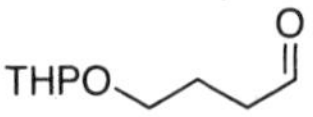 THPO 11	5	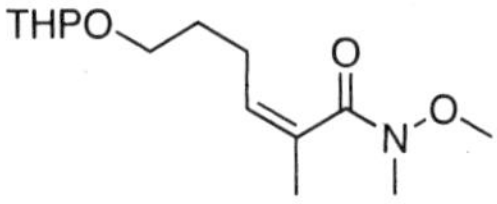 THPO 16	83
5	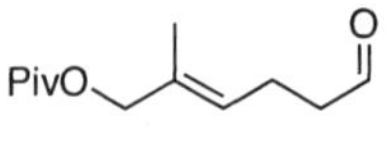 PivO 12	5	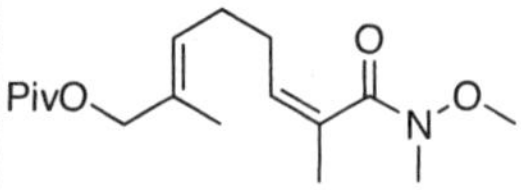 PivO 17	81

1. Department of Chemistry, University of Pennsylvania, Philadelphia, PA 19104.
2. For leading references on the preparation of ethylphosphonic acid bis(2,2,2-trifluoroethyl) ester (**2**), see: Patois, C.; Savignac, P.; About-Jaudet, E.; Collignon, N. *Synth. Commun.* **1991**, *21*(22), 2391–2396. Patois, C.; Savignac, P.; About-Jaudet, E.; Collignon, N. *Org. Synth.* **1998**, *CV 9*, 88–91.
3. For leading references on the preparation of *N*-methoxy-*N*-methylcarbamoyl chloride (**4**), see: Tius, M.A.; Busch-Petersen, J.; Yamashita, M. *Tetrahedron Lett.* **1998**, *39*, 4219–4222. Murakami, M.; Hoshino, Y.; Ito, H.; Ito, Y. *Chem. Lett.* **1998**, *2*, 163–164.

4. For leading references on the preparation of [1-(methoxymethylcarbamoyl)ethyl]phosphonic acid bis(2,2,2-trifluoroethyl) ester **5**, and its use in the synthesis of *Z*-unsaturated *N*-methoxy-*N*-methylamides, see: Fortin, S.; Dupont, F.; Deslongchamps, P. *J. Org. Chem.* **2002**, *67*, 5437–5439.
5. Nahm, S.; Weinreb, S.M. *Tetrahedron Lett.* **1981**, *22*, 3815–3818.
6. Tius, M.A.; Busch-Petersen, J. *Synlett* **1997**, 531–532.
7. Ando, T.; Yamawaki, J.; Kawate, T.; Sumi, S.; Hanafusa, T. *Bull. Chem. Soc. Jpn.* **1982**, *55*, 2504–2507.
8. Ando, T.; Kawate, T.; Yamawaki, J.; Hanafusa, T. *Chem. Lett.* **1982**, 935–938.
9. Ando, T.; Kawate, T.; Ichihara, J.; Hanafusa, T. *Chem. Lett.* **1984**, 725-728.
10. Still, W.C.; Gennari, C. *Tetrahedron Lett.* **1983**, *24*(41), 4405–4408.

Appendix
Chemical Abstracts Nomenclature (Collective Index Number); (Registry Number)

n-Butyllithium: Lithium, butyl- (8 ,9); (109-72-8)

Ethanamine, *N,N*-diethyl- (9); (121-44-8)

Ethanol, 2,2,2-trifluoro- (6,8,9); (75-89-8)

Ethylphosphonic acid bis(2,2,2-trifluoroethyl) ester: Phosphonic acid, ethyl-, bis(2,2,2-trifluoroethyl) ester (6, 8, 9); (650-16-8)

1,1,1,3,3,3-Hexamethyldisilazane: Silanamine, 1,1,1-trimethyl-*N*-(trimethylsilyl)- (9); (999-97-3)

Methanamine, *N*-methoxy-, hydrochloride (9); (6638-79-5)

Triphosgene: Methanol, trichloro-, carbonate (2:1) (9); (32315-10-9)

N-Methoxy-*N*-methylcarbamoyl chloride: Carbamic chloride, methoxymethyl- (9); (30289-28-2)

[1-(Methoxymethylcarbamoyl)ethyl]phosphonic acid bis(2,2,2-trifluoroethyl) ester: Phosphonic acid, [2-(methoxymethylamino)-1-methyl-2-oxoethyl]-, bis(2,2,2-trifluoroethyl) ester (9); (448219-33-8)

Phosphonic dichloride, ethyl- (6,7,8,9); (1066-50-8)

Pyridine (6,7,8,9); (110-86-1)

(R_S)-(+)-2-METHYL-2-PROPANESULFINAMIDE

[*tert*-Butylsulfinamide]

A. EtOH

ligand

B. ligand, $VO(acac)_2$ acetone, 0 °C; 30% H_2O_2 (aq)

C. a) $LiNH_2$, NH_3, THF, $Fe(NO_3)_3{\cdot}9H_2O$; b) ice, $ClCH_2CO_2H$

Submitted by Daniel J. Weix and Jonathan A. Ellman.[1]
Checked by Xiao Wang and Dennis P. Curran.

1. Procedures

A. *(1S,2R)-1-[(2-Hydroxy-3,5-di-tert-butylbenzylidene)amino]indan-2-ol* (Note 1). A 100-mL round-bottomed flask equipped with a magnetic stir bar is charged with (1*S*,2*R*)-(−)-*cis*-1-amino-2-indanol (0.77 g, 5.17 mmol) (Note 2), 3,5-di-*tert*-butyl salicylaldehyde (1.21 g, 5.16 mmol), and 33 mL of absolute ethanol. The yellow reaction mixture is stirred under argon for 2 hr, and the solvent is removed under reduced pressure to give a yellow oil. The oil is dissolved in 33 mL of dichloromethane and the solvent is again removed under reduced pressure. The dissolution and evaporation are repeated. After careful vacuum drying (the substance foams as the vacuum is initially applied), the resulting solid (1.85 g, 98%) is crushed to a yellow powder (Note 3) that is used without purification in Step B. *CAUTION: 30% hydrogen peroxide can cause burns.*

B. *(R_S)-(+)-tert-Butyl tert-butanethiosulfinate.* A three-necked, 1-L round-bottomed flask fitted with an overhead stirrer and two rubber septa is charged with the ligand from *Step A* (1.85 g, 5.06 mmol) and vanadyl acetylacetonate (1.33 g, 5.00 mmol) (Note 4). Acetone (250 mL) is added and the dark green reaction mixture is stirred vigorously for 30 min while open to the air. Di-*tert*-butyl disulfide (178 g, 1.00 mol) (Note 5) is slowly poured in from a graduated cylinder, and the resulting mixture is cooled to 0°C (Note 6). The reaction mixture is stirred vigorously and 30% aq hydrogen peroxide (110 mL, 1.10 mol) (Notes 7 and 8) is added over 20 hr by using two syringe pumps (Note 9). Upon the addition of the first few drops of hydrogen peroxide, the color of the mixture changes from dark green to purple, and the purple color deepens during the addition. Reaction progress can be monitored by ^{1}H NMR spectroscopy (Note 10). After 20 h, the reaction is quenched at 0°C by adding 50 mL of saturated aqueous sodium thiosulfate by syringe pump over 30 min (Note 11). The purple color fades through red to light blue-green. The resulting mixture is transferred to a 2-L separatory funnel. The reaction flask is rinsed with hexanes (250 mL) and the hexanes wash is added to the separatory funnel. The mixture is shaken, the layers are separated and the aqueous layer is extracted with hexanes (2 x 250 mL). The combined organic layers are then washed with brine (50 mL), dried over sodium sulfate and filtered. The solvent is removed under reduced pressure at 30°C or lower. The resulting yellow or brown oil (216-231 g) (Notes 10 and 12) is used without purification for *part C* (Notes 13 and 14).

CAUTION: Ammonia is toxic and corrosive and should be used only in a well-ventilated fume hood.

CAUTION: This procedure generates hydrogen gas and should be performed in a well-ventilated fume hood.

CAUTION: tert-Butyl thiol, although efficiently scavenged, is produced in stoichiometric quantities. It has a foul odor and is the odorant used for natural gas.

C. *(R_S)-(+)-tert-Butylsulfinamide.* A 5-L, three-necked, round-bottomed flask equipped with a mechanical stirrer (glass rod and glass paddle), a cold finger fitted with a gas inlet, and a Claisen adapter fitted with a glass stopper and a gas outlet connected to a mineral oil bubbler is placed under argon and cooled in a bucket of dry ice/acetone. The cold finger is filled with dry ice/acetone, and the gas inlet is changed from argon to an ammonia cylinder (Note 15). The ammonia gas pressure is adjusted so that most of the ammonia is condensed into the vessel as it passes through the cold finger (Note 16). After about 2 L of ammonia is condensed (about 2.5 h), the mineral oil bubbler is removed from the Claisen adapter gas outlet and replaced with an argon line. The ammonia line is removed from the cold finger gas inlet and replaced with a mineral oil bubbler.

The reaction mixture at –75°C (internal temperature) is flushed gently with an argon stream, and $Fe(NO_3)_3 \cdot 9H_2O$ (1.0 g, 2.48 mmol) (Note 17) is added. The mixture turns brown as dissolution occurs on stirring. The Claisen adaptor is replaced with an adaptor containing a low temperature thermometer. The cooling bath is lowered and pieces of lithium ribbon (17.4 g, 2.50 mol) (Note 18) are added with tweezers in ~1 g portions (Note 19). Each piece is cut in half just prior to addition to ensure fast dissolution and reaction. As each piece of lithium is added, the mixture briefly turns a dark blue color. This fades to give a gray suspension of lithium amide with concomitant evolution of hydrogen gas (Note 19). Once the reaction returns to the gray color, another piece of lithium is added. The dry ice/acetone bath is periodically raised to contact the bottom of the flask to slow the refluxing caused by the formation of $LiNH_2$ (Note 19). After the addition of the lithium is complete (about 1.5 hr), the reaction mixture is allowed to reflux for 15 min (Note 20), and then submerged in the dry ice/acetone bath.

After cooling to –78°C (about 30 minutes) and replacement of the thermometer adaptor by an addition funnel, a solution of crude (R_S)-(+)-*tert*-butyl *tert*-butylthiosulfinate (entire crude product from Step B, 1.00 mol) in 320 mL of freshly distilled, dry THF is added to the vigorously stirring reaction mixture over 45 minutes via the addition funnel (Note 21). The addition funnel is rinsed with a small amount (5-10 mL) of THF and this is added to the reaction mixture. The addition funnel is replaced by the thermometer adaptor. After one hour at –78 to –72°C, the cooling bath is removed, stirring is slowed to the lowest possible rate and the reaction mixture is allowed to slowly warm to room temperature overnight (15 hr) under a stream of argon (Note 22).

The resulting thick gray mixture, still under an argon stream, is cooled in an ice-water bath. Ice (540 g, 30 mmol) is added in one portion, and the dark green mixture is stirred until it is mostly homogeneous (Note 23). Chloroacetic acid (104 g, 1.10 mol) (Note 24) is added to the ice-cold mixture in five portions over 20 min. The resulting mixture is stirred overnight (12 hr) at ambient temperature under argon. The claret solution is poured into a 2 L separatory funnel and extracted with methylene chloride (6 x 1 L). The combined organic layers are dried over Na_2SO_4, and filtered through a pad of celite. The solvent is removed under reduced pressure.

Hexanes (200 mL) are added and the solvent is removed under reduced pressure. More hexanes (50 mL) are added and the evaporation is repeated. The resulting yellow-orange solid is triturated (Note 25) with hexanes (250 mL) in a 500-mL Erlenmeyer flask. After the solids are crushed, a magnetic stir bar is added and the slurry is stirred for 30 minutes. The solids are collected on a suction filter, washed with hexanes (15 mL), and suction-dried (Note 26). The triturated material is then recrystallized by dissolving it in boiling hexanes (150 mL), followed by slow cooling to room temperature with fast stirring. The crystals are collected by suction filtration and washed with cold hexanes (300 mL, precooled to –30°C) and suction-dried to provide 84.9-86.9 g (70-72% from di-*tert*-butyl disulfide) of the desired product (Note 27).

2. Notes

1. This ligand has been reported several times.[2] This procedure is adapted from the procedure of Ruck and Jacobsen.[2d] (1*S*,2*R*)-(–)-*cis*-1-Amino-2-indanol leads to (R_S)-*tert*-butanesulfinamide while (1*R*,2*S*)-(+)-*cis*-1-amino-2-indanol leads to (S_S)-*tert*-butanesulfinamide.

2. (1*S*,2*R*)-(–)-*cis*-1-Amino-2-indanol was purchased from Strem (submitters) or Aldrich (checkers). 3,5-Di-*tert*-butyl-2-hydroxybenzaldehyde was purchased from Aldrich.

3. The yellow powder exhibited the following features: mp 64 - 68°C (65 – 68°C lit.[2c]); ^{1}H NMR ($CDCl_3$, 300 MHz) δ: 1.33 (s, 9 H), 1.43 (s, 9 H), 2.17 (br s, 1 H), 3.13 (dd, J = 4.9, 15.9 Hz, 1 H), 3.26 (dd, J = 5.8, 15.9 Hz, 1 H), 4.69 (m, 1 H), 4.81 (d, J = 5.3 Hz, 1 H), 7.19 (d, J = 2.2 Hz, 1 H), 7.20-7.32 (m, 4 H), 7.44 (d, J = 2.3 Hz, 1 H), 8.63 (s, 1 H), 13.12 (br s, 1 H); ^{13}C NMR ($CDCl_3$, 75 MHz) δ: 29.4, 31.5, 34.1, 35.0, 39.7, 75.2, 75.7, 117.8, 124.9, 125.4, 126.5, 127.0, 127.6, 128.5, 136.9, 140.4, 140.8, 140.9, 157.9, 168.3; IR (neat) 739, 755, 1626, 2957 cm^{-1}; HRMS-FAB (*m/z*) calcd for $C_{24}H_{31}NO_2$, 365.2355; Found, 365.2353. Anal. Calcd for $C_{24}H_{31}NO_2$: C, 78.86; H, 8.55; N, 3.83. Found: C, 79.15; H, 8.69; N, 3.69.

4. Vanadium (IV) bis(acetylacetonato)oxide was purchased from Strem (submitters) or Aldrich (checkers). Industrial Grade acetone was used as received.

5. Di-*tert*-butyl disulfide was purchased from Acros and distilled (91-96°C/10 mmHg). The distillation proceeds smoothly if the pot is not allowed to boil vigorously, otherwise excessive foaming and bumping can result. Avoid Boileezers® or stir bars in the pot.

6. A recirculating chiller (Lauda RM-6B or FTS System) with an external cooling coil and a 2-propanol bath was used. Slightly higher temperatures result in a small decrease in enantioselectivity, but no decrease in conversion.

7. Hydrogen peroxide (30% aq, stabilized with sodium stannate) was purchased from Fisher Scientific.

8. This mixture of hydrogen peroxide, acetone, and water is considered safe. See the "hydrogen peroxide product information manual" available from Eka chemicals (www.eka.com).

9. The submitters used a model NE-1600 programmable six-syringe pump from New Era Pump Systems, Inc., while checkers used two separate syringe pumps, model M361 from Thermo Orion.

10. ^{1}H NMR ($CDCl_3$, 300 MHz) δ: 1.39 (s, 9 H), 1.54 (s, 9 H). When a reaction is incomplete, a starting material resonance, is visible at 1.31 ppm; ^{13}C NMR ($CDCl_3$, 75 MHz) δ: 24.2, 32.3, 48.6, 59.3. Analysis by chiral phase HPLC (Daicel Chiralpak AS-H column, 25 cm x 0.46 cm; 97:3 hexanes:2-propanol; 1 mL/min; 254 nm; (*S*) R_t = 6.6 min; (*R*) R_t = 7.8 min) shows that the product has an 86-87% ee.

11. It is best to quench any remaining hydrogen peroxide before allowing the reaction to warm to room temperature because a mild exotherm can occur due to hydrogen peroxide decomposition upon warming. When saturated aq sodium thiosulfate was added in one portion with cooling, an exotherm occurred which raised the internal temperature of the reaction to 32°C; therefore, slow addition with adequate cooling is preferred.

12. By ^{1}H NMR spectroscopic analysis, the product contained less than 1% acetone (peak at 2.17 ppm in $CDCl_3$). If acetone is still present, it is removed by adding more hexanes and removing the azeotrope under reduced pressure. The checkers sometimes observed small amounts of an insoluble yellow solid in the late stages of solvent removal. This is elemental sulfur (mp 111-113°C, lit. 112.8°C), and was removed by decantation. The decomposition of $S_2O_3^{-2}$ to S and SO_3^{-2} is a well-known reaction.

13. *tert*-Butyl *tert*-butylthiosulfinate is a useful starting material for the synthesis of *tert*-butyl sulfoxides. It can be purified by distillation or crystallization if needed.[2a]

14. The thiosulfinate should be stored at –20°C. At this temperature it is a solid and no decomposition has been observed over several months. Gradual decomposition and erosion of enantiopurity occur on storage at room temperature.

15. Ammonia gas (99.99% pure) was purchased from Matheson or Praxair.

16. The dry ice evaporates rapidly from the cold finger if the ammonia is passed in too quickly.

17. $Fe(NO_3)_3{\bullet}9H_2O$ was purchased from Mallinckrodt (submitters) or Aldrich (checkers).

18. The submitters used lithium bars (12.7 mm x 165 mm, 99.9 % pure) purchased from Alfa Aesar. The checkers used lithium ribbon (dissolves more rapidly) purchased from Aldrich (45 mm x 0.75 mm, 99.9%). The lithium was washed with hexanes to remove storage oil prior to rapid weighing in a tared beaker of hexanes. The thermometer adaptor was partially removed to drop in the lithium.

19. Lithium amide does not form rapidly at –78°C, but the blue color begins to form as the mixture is warmed to about –45°C. During the addition, the cooling bath is raised and lowered periodically to keep the internal temperature between –50°C and –45°C.

20. All the lithium must be dissolved and converted to lithium amide to prevent an exotherm from occurring during the subsequent evaporation of the ammonia.

21. The internal temperature is kept at –73°C or below. The thiosulfinate is added directly into the reaction mixture to prevent it from solidifying on the sidewalls of the vessel. Adjusting the mixing speed to wash any solidified material back into the reaction mixture may be necessary.

22. The submitters report that stirring is not needed during this stage.

23. The initially thick precipitate largely dissolves, but a small amount of powder remains suspended. If the mixture is not stirred during the ammonia evaporation or if the stirring stops, it can be difficult to restart the stirrer. Churning the glass rod by hand breaks up the mass so that stirring can be reinitiated.

24. Chloroacetic acid was purchased from Acros or Aldrich.

25. Trituration improves the enantiopurity and removes most of the colored impurities. The recovery in the trituration will decrease if any CH_2Cl_2 remains in the crude material because *tert*-butylsulfinamide is very soluble in this solvent.

26. This material has a 95% ee and represents a 75% yield from di-*tert*-butyl disulfide.

27. The product is a white to off-white crystalline solid; mp 102 - 105°C (102 - 103°C lit.[2a]); ^1H NMR ($CDCl_3$, 300 MHz) δ: 1.22 (s, 9 H), 3.73 (s, 2 H). ^{13}C NMR ($CDCl_3$, 75 MHz) δ: 22.1, 55.2; IR (neat) 1030, 1363, 1460, 1588, 3119, 3216 cm^{-1}; Anal. Calcd for $C_4H_{11}NOS$: C, 39.64; H, 9.15; N, 11.56. Found: C, 39.62; H, 9.12; N, 11.29. Chiral phase HPLC (Daicel

Chiralpak AS-H column, 25 cm x 0.46 cm; 90:10 hexanes:ethanol; 1.2 mL/min; 220 nm; (*R*) r_t = 7.9 min; (*S*) r_t = 10.6 min) shows that the product has ≥ 99% ee; The submitters report that the analysis can also be conducted on a Daicel Chiralpak OD column; 93/7, hexanes:i-PrOH, 1 mL/min, 220 nm; (*S*) R_t = 23.7 min, (*R*) R_t = 27.7 min;

Waste Disposal Information

All hazardous materials should be handled and disposed of in accordance with "Prudent Practices in the Laboratory"; National Academy Press; Washington, DC, 1995.

3. Discussion

Chiral amines are key components of many pharmaceutical agents, materials, and catalysts. Since its introduction in 1997,[3] enantiopure *tert*-butylsulfinamide has proven to be an extremely versatile chiral ammonia equivalent for the asymmetric synthesis of amines.[4]

The original synthesis included an oxidation that did not scale well and did not provide procedures for removing the offensive *tert*-butyl thiol odor. Recent work has improved the scalability of the oxidation, allowed easy access to both (R_S)- and (S_S)-*tert*-butylsulfinamide, and eliminated the tedious distillation of the *tert*-butyl *tert*-butylthiosulfinate intermediate.[2g] This procedure used sodium hydroxide solutions to trap the thiol; however, the thiol removal was incomplete. *tert*-Butyl thiol is the primary odorant added to natural gas because its odor can be detected at very low levels, and because it is heavier than air and resists oxidation well.

We initially explored trapping the thiol by the use of gas washing bottles filled with bleach solution. While this was effective at controlling odor, it presents an unnecessary danger when combined with ammonia (possible formation of chlorine gas, nitrogen trichloride, or hydrazine; explosion hazard). After a survey of various methods, we found that evaporation of the ammonia before quenching the reaction with ice eliminated thiol odor during the ammonia removal process. Furthermore, if chloroacetic acid is then added to the crude aqueous solution and the mixture stirred, the residual lithium *tert*-butyl thiolate reacts with the lithium chloroacetate to form the highly water soluble lithium (*tert*-butylthio)acetate. At this stage, the reaction mixture has no discernible thiol odor. Simple extraction with methylene chloride then separates the sulfinamide from the lithium (*tert*-butylthio)acetate. Trituration followed by crystallization yields highly enantioenriched *tert*-butylsulfinamide. This procedure allows for the preparation of either enantiomer of *tert*-butylsulfinamide in good yield on mole scale with no thiol odor and only trituration and crystallization for purification.

1. Department of Chemistry, University of California, Berkeley, CA 94720; jellman@uclink.berkeley.edu.
2. (a) Cogan, D. A.; Liu, G.; Kim, K.; Backes, B. J.; Ellman J. A. *J. Am. Chem. Soc.* **1998**, *120*, 8011-8019. (b) Zhen, L.; Fernandez, M.; Jacobsen, E. N. *Org. Lett.* **1999**, *1*, 1611-1613. (c) Flores-Lopez, L. Z.; Parra-Hake, M.; Somanathan, R.; Walsh, P. J. *Organometallics* **2000**, *19*, 2153-2160. (d) Ruck, R. T.; Jacobsen, E. N. *J. Am. Chem. Soc.* **2002**, *124*, 2882-2883. (e) Gama, A.; Flores-Lopez, L. Z.; Aguirre, G.; Parra-Hake, M.; Somanathan, R.; Walsh, P. J. *Tetrahedron Asymm.* **2002**, *13*, 149-154. (f) Beatrice, P.; Anson, M. S.; Campbell, I. B.; MacDonald, S. J. F.; Priem, G.; Jackson, R. F. W. *Synlett* **2002**, 1055-1060. (g) Weix, D. J.; Ellman, J. A. *Org. Lett.* **2003**, *5*, 1317-1320.
3. Liu, G.; Cogan, D.; Ellman, J. A. *J. Am. Chem. Soc.* **1997**, *119*, 9913-9914.
4. (a) Ellman, J. A.; Owens, T. D.; Tang, T. P. *Acc. Chem. Res.* **2002**, *35*, 984-995. (b) Ellman, J. A. *Pure Appl. Chem.* **2003**, *75*, 39-46. (c) Mukade, T.; Dragoli, D. R.; Ellman, J. A. *J. Comb. Chem.* **2003**, 590-596. (d) Kochi, T.; Tang, T. P.; Ellman, J. A. *J. Am. Chem. Soc.* **2003**, 11276-11282. (e) Evans, J. W.; Ellman, J. A. *J. Org. Chem.* **2003**, 9948-9957. (f) Kells, K. W.; Chong, J. M. *Org. Lett.* **2003**, *5*, 4215-4218. (g) Cooper, I. R.; Grigg, R.; Hardie, M. J.; MacLachlan, W. S.; Sridharan, V.; Thomas, W. A. *Tetrahedron Lett.* **2003**, *44*, 2283-2285 (h) Rech, J. R.; Floreancig, P. E. *Org. Lett.* **2003**, *5*, 1495-1498. (i) Jung, P. M.; Beaudegnies, R.; De Mesmaeker, A.; Wendeborn, S. *Tetrahedron Lett.* **2003**, *44*, 293-297.

Appendix
Chemical Abstracts Nomenclature (Registry Number)

(1*S*,2*R*)-(−)-*cis*-1-Amino-2-indanol: 1*H*-Inden-2-ol, 1-amino-2,3-dihydro-, (1*S*,2*R*)-; (126456-43-7)

3,5-Di-*tert*-butyl salicylaldehyde: Benzaldehyde, 3,5-bis(1,1-dimethylethyl)-2-hydroxy-; (37942-07-7)

(1*S*,2*R*)-1-[(2-Hydroxy-3,5-di-*tert*-butylbenzylidene)amino]indan-2-ol: 1*H*-Inden-2-ol, 1-[[[3,5-bis(1,1-dimethylethyl)-2-hydroxyphenyl]-methylene]amino]-2,3-dihydro-, (1*S*,2*R*)-; (212378-89-7)

(R_S)-(+)-*tert*-Butyl *tert*-butylthiosulfinate: 2-Propanesulfinothioic acid, 2-methyl-, *S*-(1,1-dimethylethyl) ester, [*S*(*R*)]-; (67734-35-4)

Vanadyl acetylacetonate: Vanadium, oxobis(2,4-pentanedionato-**κ***O*,***k****O*')-; (3153-26-2)

Di-*tert*-butyl disulfide: Disulfide, bis(1,1-dimethylethyl); (110-06-5)

Hydrogen peroxide; (7722-84-1)

Sodium thiosulfate: Thiosulfuric acid, disodium salt; (7772-98-7)

Ammonia; (7664-41-7)

$Fe(NO_3)_3 \cdot 9H_2O$ (Iron(III) nitrate nonahydrate): Nitric acid, iron(3+) salt, nonahydrate; (7782-61-8)

Lithium; (7439-93-2)

Chloroacetic acid; (79-11-8)

(R_S)-(+)-2-Methyl-2-Propanesulfinamide [*tert*-Butylsulfinamide]: 2-Propanesulfinamide, 2-Methyl-, [*S*(*R*)]-; (196929-78-9)

PREPARATION OF 1-BUTYL-3-METHYLIMIDAZOLIUM TETRAFLUOROBORATE

[1*H*-Imidazolium, 1-butyl-3-methyl, tetrafluoroborate (1–)]

$$H_3C-N^{+}\!\!\diagdown\!N-C_4H_9\ Cl^{-} \xrightarrow[H_2O]{NaBF_4} H_3C-N^{+}\!\!\diagdown\!N-C_4H_9\ BF_4^{-}$$

Submitted by Xavier Creary and Elizabeth D. Willis.[1]
Checked by Gustavo Moura-Letts and Dennis P. Curran.

1. Procedure

1-Butyl-3-methylimidazolium tetrafluoroborate. 1-Butyl-3-methylimidazolium chloride (30.00 g, 172 mmol) (Note 1) is placed in a 125-mL Erlenmeyer flask containing a stir bar and a thermometer. This salt is dissolved in 35 mL of distilled water and $NaBF_4$ (20.00 g, 182 mmol) (Note 2) is added in portions with stirring over 10-15 min. The $NaBF_4$ dissolves as the mixture emulsifies and cools to 14°C (Note 3). After the mixture warms back to ambient temperature, 30 mL of CH_2Cl_2 is added and the contents are transferred to a 125-mL separatory funnel. The bottom CH_2Cl_2 phase is separated (Note 4). The aqueous phase is extracted with an additional 20 mL of CH_2Cl_2. The combined CH_2Cl_2 phases are shaken in a separatory funnel with a solution of $NaBF_4$ (10.0 g, 91 mmol) in 20 mL of water. The CH_2Cl_2 phase is separated and dried over a mixture of 1.0 g of Na_2SO_4 and 3.0 g of $MgSO_4$. The mixture is filtered through a Büchner funnel and the salts are washed with an additional 15 mL of CH_2Cl_2. A short path distillation head is attached to the 250-mL round-bottomed flask containing the filtrate and most of the CH_2Cl_2 is removed by distillation at 30 mm pressure. Care is taken not to heat the product above 50°C. The solvent is condensed in a receiver flask cooled in an ice-water slurry (Note 5). The last traces of CH_2Cl_2 are removed using a rotary evaporator at 15 mm and 45°C followed by vacuum drying at ambient temperature until the weight remains constant. The pure 1-butyl-3-methylimidazolium tetrafluoroborate (34.2–34.3 g; 89% yield) is a colorless to pale yellow, viscous liquid (Notes 6, 7 and 8).

2. Notes

1. The submitters prepared 1-butyl-3-methylimidazolium chloride as previously described.[2] The checkers purchased this from Aldrich Chemical Co.

2. Sodium tetrafluoroborate (98%) was purchased from Acros Organics and used as received.

3. If the emulsion is allowed to stand for 30-60 min, then two phases results. However, there is no need to do this since addition of CH_2Cl_2 induces rapid phase separation.

4. The product at this point is not completely chloride free. In a separate analysis by the submitters, removal of the CH_2Cl_2 from this solution and analysis of 1.53 g of product by titration with 0.100 M $AgNO_3$ (Mohr titration; K_2CrO_4 indicator)[3] required 0.6 mL of the silver nitrate solution to precipitate all of the chloride ion. This corresponds to a product containing 0.7% $BMIM^+ Cl^-$ and 99.3% $BMIM^+ BF_4^-$.

5. The distilled CH_2Cl_2 has droplets of water from the azeotrope with CH_2Cl_2. This distillation process helps to dry the product and allows recovery of most of the CH_2Cl_2 used in the procedure.

6. Dissolution of 1.50 g of this liquid in 3 mL of water followed by addition of 0.100 M $AgNO_3$ gave no precipitation or cloudiness.

7. The submitters report that the water content of the product was 0.17% as determined from the ^1H NMR spectrum of the neat liquid by integration of the water signal at δ 2.9 and the CH_2 signal at δ 1.93. The product is hygroscopic and water content increases with exposure to air. The checkers could not detect water in their spectrum in CD_2Cl_2.

8. The product exhibits the following spectroscopic properties: IR (thin film) 3646, 3162, 3122, 2964, 2877, 1575, 1467, 1431, 1171, 1048, 850 cm^{-1}; ^1H NMR (500 MHz, CD_2Cl_2) δ: 0.91 (t, 3 H, J = 7.3 Hz), 1.32 (sextet, 2 H, J = 7.3 Hz), 1.81 (quintet, 2 H, J = 7.4 Hz), 3.89 (s, 3 H), 4.14 (t, 2 H, J = 7.3 Hz), 8.66 (s, 1 H), 7.37 (s, 2 H); ^{13}C NMR (125 MHz, CD_2Cl_2) δ: 13.1, 19.3, 31.9, 36.1, 49.7, 122.5, 123.8, 136.2. Anal. Calcd for $C_8H_{15}BF_4N_2$: C, 42.51; H, 6.69; N, 12.37; Found: C, 41.94; H, 6.85; N, 12.37; trace analysis for Cl, 636 ppm.

Safety and Waste Disposal Information

All hazardous materials should be handled and disposed of in accordance with "Prudent Practices in the Laboratory"; National Academy Press; Washington, DC, 1995.

3. Discussion

This procedure is based on the general method recently reported in *Organic Syntheses*.[2] The preparation of 1-butyl-3-methylimidazolium hexafluorophosphate ($BMIM^+ PF_6^-$) proceeds as described.[2] However, in our hands, the reported method for the preparation of 1-butyl-3-methylimidazolium tetrafluoroborate ($BMIM^+ BF_4^-$; the most commonly used of the ionic liquids) was not reproducible. The product was contaminated with varying (and significant) amounts of $BMIM^+ Cl^-$[4]. The use of the relatively insoluble KBF_4 (0.44 g/100 mL) in the original preparation makes dissolution and subsequent reaction problematic. Thus, analysis of 0.716 g of a typical ionic liquid product produced from KBF_4 by titration with 0.100 M $AgNO_3$ required 11.1 mL of the silver nitrate solution to precipitate all of the chloride ion. This corresponds to a product containing 27% $BMIM^+ Cl^-$ and 73% $BMIM^+ BF_4^-$.

The procedure described herein uses the much more soluble $NaBF_4$ (97.3 g/100 mL). After the first cycle, the chloride content is only 0.7%, and the chloride can no longer be detected after the second cycle. In addition, the tedious process of removing water by distillation under reduced pressure is eliminated. This procedure is also useful for the preparation of the ionic liquid *n*-butyl pyridinium tetrafluoroborate, which is subject to the same chloride contamination problems when prepared from KBF_4. Finally, the amount of residual water in the product can be readily determined from the ^1H NMR spectrum of the neat product.

1. Department of Chemistry and Biochemistry, University of Notre Dame, Notre Dame, IN 46556.
2. Dupont, J.; Consorti, C. S.; Suarez, P. A. Z.; de Sousa, R. F. *Organic Syntheses* **2002** *79*, 236.
3. Christian, G. D. "Analytical Chemistry, Fifth Edition", John Wiley and Sons, Inc., New York, 1994, p. 278.
4. Contamination of $BMIM^+ BF_4^-$ and other ionic liquids with chloride in various preparations has been noted in the literature. See Seddon, K. R.; Stark, A.; Torres, M.-J. *Pure Appl. Chem.* **2000**, *72*, 2275.

Appendix
Chemical Abstracts Nomenclature; (Registry Number)

1-Butyl-3-methylimidazolium chloride: 1*H*-Imidazolium, 1-butyl-3-methyl-, chloride; (79917-90-1)

$NaBF_4$: Borate(1-), tetrafluoro-, sodium; (13755-29-8)

1-Butyl-3-methylimidazolium tetrafluoroborate: 1*H*-Imidazolium, 1-butyl-3-methyl, tetrafluoroborate (1–); (174501-65-6)

CATALYTIC ASYMMETRIC ACYL HALIDE-ALDEHYDE CYCLOCONDENSATION REACTION

[(4*S*)-4-(2-Phenylethyl)-2-oxetanone]

A. H_2N / OH, *i*-Pr → $(CF_3SO_2)_2O$, Et_3N; CH_2Cl_2, -78 °C → SO_2CF_3 N aziridine, *i*-Pr **1**

B. **1** → $BnNH_2$; 100 °C → **2** (*i*-Pr, Bn N, *i*-Pr, F_3CO_2S NH, HN SO_2CF_3)

C. **2** → Me_3Al; CH_2Cl_2 → **3** (*i*-Pr, Bn N, *i*-Pr, F_3CO_2S N—Al—N SO_2CF_3, Me)

Me(C=O)Br + H(C=O)CH₂CH₂Ph → 10 mol% **3**; *i*-Pr_2NEt, CH_2Cl_2; -50 °C → oxetanone (Ph)

Submitted by Scott G. Nelson and Paul M. Mills.[1]
Checked by Takashi Ohshima and Masakatsu Shibasaki.

1. Procedure

A. *(S)-N-Trifluoromethylsulfonyl-2-isopropylaziridine.* An oven-dried, 500-mL, round-bottomed flask is equipped with Teflon-coated magnetic stirring bar and sealed with a rubber septum containing a needle adapter to a N_2 source. The flask is charged with (*S*)-valinol (8.30 g, 80.5 mmol) (Note 1), triethylamine (26.3 mL, 189 mmol) (Note 2) and 135 mL of dichloromethane (Note 3). The flask is placed in a –78°C dry ice-acetone bath (Note 4) and, to the well-stirred solution, is added of trifluoromethanesulfonic anhydride (31.8 mL, 189 mmol) (Note 5) via syringe over 20 min (Note 6). The resulting reaction mixture is held at –78°C for 5 hr. The reaction mixture is transferred to a 500-mL separatory funnel containing 200 mL of 0.1 N HCl and the mixture is thoroughly shaken and the layers are separated. The organic portion is washed successively with one portion of 0.1 M HCl (200 mL), two portions of saturated aqueous $NaHCO_3$ (200 mL each) and one portion of brine (200 mL). The organic portion is dried over anhydrous magnesium sulfate (Note 7), filtered and concentrated under reduced pressure on a rotary evaporator (Note 8) to afford 16.8-16.9 g (96-97%) of (*S*)-*N*-trifluoromethylsulfonyl-2-isopropylaziridine (**1**) as a pale yellow oil (Note 9). The (*S*)-*N*-trifluoromethylsulfonyl-2-isopropylaziridine is used in the next transformation without further purification.

B. *(2S,6S)-4-Benzyl-1,7-bis(trifluoromethylsulfonyl)-2,6-diisopropyl-1,4,7-triazaheptane.* A flame-dried, 250-mL round-bottomed flask containing a Teflon-coated magnetic stirring bar is charged with (*S*)-*N*-trifluoromethylsulfonyl-2-isopropylaziridine (**1**) (16.9 g, 77.8 mmol). Benzylamine (4.03 mL, 37.0 mmol) (Note 10) is added resulting in a mild exotherm (Note 11). Once the exotherm subsided (approx. 10 min), the reaction flask is placed in a 100°C oil bath and held at this temperature for 12 hr. The crude product mixture is separated by flash chromatography on silica gel (approx. 300 g) (Note 12) using 10% ethyl acetate in hexane as the eluent (Note 13) to give 20.4-21.4 g of a pale yellow, sticky solid, which included 18.1-18.5 g of **2** (90-92% yield based on ^{1}H NMR analysis), ethyl acetate, and trace amount of yellow material (Note 14). This sticky solid is further purified by trituration with hexane (20 mL, vigorous stirring) and after filtration 15.6-16.0 g (78-80%) of the triamide **2** is obtained as a white powder (Note 15).

C. *Acyl halide-aldehyde cyclocondensation: (4S)-4-(2-Phenethyl)-oxetan-2-one.* An oven-dried, 500-mL, three necked, round-bottomed flask is equipped with a Teflon-coated magnetic stirring bar and a 125-mL addition funnel and is sealed with rubber septa containing a needle adapter to a N_2 source and a Teflon-coated thermocouple probe attached to a digital thermometer. The flask is charged with triamine **2** (4.00 g, 7.40 mmol) and dichloromethane (150 mL), whereupon a 2M hexanes solution of trimethylaluminum (3.70 mL) (Note 16) is added at ambient temperature (23°C) (*CAUTION*: *Methane gas evolution*). The resulting solution is stirred at 23°C for 2 hr before being cooled to –50°C (Note 17). Once the reaction mixture has reached –50°C, diisopropylethylamine (22.0 mL, 126 mmol)

(Note 18) and acetyl bromide (10.4 mL, 141 mmol) (Note 19) are added consecutively via syringe at a rate that maintains the internal reaction temperature ≤ –42°C. The addition funnel is charged with hydrocinnamaldehyde (9.75 mL, 74.0 mmol) and CH_2Cl_2 (10 mL) and this solution is added dropwise to the reaction mixture at a rate that maintains the internal temperature at ≤ –46°C (Note 20). Once addition is complete, the reaction mixture is stirred at –50°C for 16 h. The reaction mixture is diluted with 350 mL CH_2Cl_2, then transferred to a 2-L separatory funnel containing 800 mL of 0.1 M HCl and the mixture is thoroughly shaken and the layers are separated. The organic portion is washed consecutively with two portions of 0.1 M HCl (800 mL each), three portions of saturated aqueous $NaHCO_3$ (800 mL each) and two portions of brine (800 mL each). After each washing, the separated aqueous payer is extracted with one portion of diethyl ether (400 mL) using the same portion of ether for each extraction; the ethereal extract is reserved until washing of the CH_2Cl_2 layer is complete. The CH_2Cl_2 solution and the ether extract are combined, dried over anhydrous sodium sulfate (Note 21), filtered and concentrated under reduced pressure on a rotary evaporator. The crude product mixture is distilled under reduced pressure (100°C at 5 Pa using turbo-molecular pump) (Note 22) yielding 10.4 g (80%) of (4*S*)-4-(2-phenethyl)oxetan-2-one (Notes 23 and 24).

2. Notes

1. The submitters prepared (*S*)-valinol according to an *Organic Syntheses* procedure employing 200 g (1.71 mol) of (*S*)-valine, 100 g (2.64 mol) of lithium aluminum hydride and 6 L of THF; the THF is used directly from a freshly opened bottle and is not distilled. The crude product mixture was purified by vacuum distillation (80°C at 7 mm Hg) to afford 118 g (67%) of (*S*)-valinol. See: Dickman, D. A.; Meyers, A. I. *Org. Synth., Coll. Vol. VII* **1990**, 530. The checker used commercially available (*S*)-valinol, which was purchased from Aldrich Chemical Company, after distillation (70 °C at 1.3 kPa).

2. Triethylamine was purchased from Fisher Scientific Company and was freshly distilled over calcium hydride.

3. Methylene chloride was purchased from EM Science and was

freshly distilled over calcium hydride.

4. Bath temperature was achieved using a dry ice-acetone slurry.

5. Trifluoromethanesulfonic anhydride was purchased from Aldrich Chemical Company and was used as received. The use of only 2 mol equivalent of trifluoromethanesulfonic anhydride (27.0 mL, 161 mmol) gave almost identical result.

6. Internal temperature was gradually increased to ca. –60°C.

7. Anhydrous magnesium sulfate was purchased from EM Science.

8. The residue was concentrated using diaphragm pump (2 kPa). Concentration using oil-pump (0.5 kPa) caused loss of the product due to the low boiling point of the product.

9. Spectral data for the crude aziridine: ^{1}H NMR (500 MHz, $CDCl_3$) δ: 1.03 (d, J = 7.0 Hz, 3 H), 1.05 (d, J = 7.0 Hz, 3 H), 1.65 (octet, J = 7.0 Hz, 1 H), 2.45 (d, J = 5.0 Hz, 1 H), 2.86 (d, J = 7.0 Hz, 1 H), 2.90 (td, J = 7.0, 5.0 Hz, 1 H).

10. Benzylamine was purchased from Aldrich Chemical Company and freshly distilled before use.

11. Internal temperature reached to 65°C even when benzylamine was added very slowly (0.5 mL/min).

12. Silica gel was purchased from Merck (Silica gel 60, 230-400 mesh ASTM).

13. The submitters used 80% hexanes-20% ethyl acetate mixture as the eluent for flash chromatography on silica gel purchased from Bodman Industries (70-239 mesh). Using this eluent system, however, the checker obtained unsatisfactory separation.

14. The submitters used the pale yellow solid for the next reaction without further purification (90% yield). To maintain the purity of the triamide **2**, the checkers used the trituration technique.

15. The analytical data are as follows: mp 118-119°C (lit. 112°C); TLC R_f = 0.50 (hexane:EtOAc = 3:1) $[\alpha]_D$ –50° (*c* 1.6, MeOH); IR (NaCl) cm^{-1}: 3308, 2968, 2880, 2840, 1435, 1371, 1229, 1195, 1148, 1025; IR (KBr) cm^{-1}: 3262, 2969, 1449, 1379, 1231, 1195, 1151, 617; ^{1}H NMR (500 MHz, $CDCl_3$) δ: 0.84 (d, J = 6.9 Hz, 6 H), 0.87 (d, J = 7.0 Hz, 6 H), 1.93-2.00 (m, 2 H), 2.50 (dd, J = 13.5, 5.5 Hz, 2 H), 2.66 (dd, J = 13.5, 8.5 Hz, 2 H), 3.39 (d, J = 13.5 Hz, 1 H), 3.56-3.61 (m, 2 H), 3.94 (d, J = 13.5 Hz, 1 H), 7.27-7.39 (m, 5 H), 5.40 (br s, 2 H); ^{13}C NMR (125 MHz, $CDCl_3$) δ:

17.2, 18.1, 29.7, 55.4, 58.56, 58.60, 119.3 (q, J_{CF} = 319 Hz), 127.7, 128.6, 129.6, 137.0; MS (EI, 70 eV): *m/z* 542 $(M+1)^+$, 540, 337, 91; HRMS *m/z* calcd for $C_{19}H_{29}F_6N_3O_4S_2$: 541.1504, found 541.1516; Elemental Anal. Calcd for $C_{19}H_{29}F_6N_3O_4S_2$: C, 42.14; H, 5.40; N, 7.76; Found: C, 42.14; H, 5.40; N, 7.76.

16. Trimethylaluminum (2.0M solution in hexanes) was purchased from Aldrich Chemical Company and was use as received.

17. Bath temperature was controlled using a circulating chiller equipped with a submersible cooling probe.

18. *N,N*-Diisopropylethylamine was purchased from Aldrich Chemical Company and was freshly distilled over calcium hydride.

19. Acetyl bromide was purchased from Aldrich Chemical Company and was freshly distilled over P_2O_5.

20. Hydrocinnamaldehyde was purchased from Aldrich Chemical Company and was freshly distilled over calcium hydride.

21. Anhydrous sodium sulfate was purchased from L. T. Baker.

22. The checkers experienced some decomposition of the β-lactone to 3-butenylbenzene during distillation, potentially caused by trace acidic impurities (c.f., *i*-$Pr_2NEt{\cdot}HBr$) remaining in the crude product mixture). Alternatively, the crude product mixture can be purified by silica-gel column chromatography (10% EtOAc in hexane) to afford the product in 90% yield.

23. The analytical data are as follows: TLC R_f = 0.33 (hexane:EtOAc = 3:1) $[\alpha]_D$ –43.1° (*c* 1.12, CH_2Cl_2) –47.4° (*c* 1.6, $CHCl_3$, 92% ee); IR (NaCl): 3085, 2987, 1828, 1545, 1455, 1135, 700 cm^{-1}; IR (neat): 3027, 2930, 1825, 1496, 1455, , 1412, 1133, 829, 750, 700 cm^{-1}; ^1H NMR (500 MHz, $CDCl_3$) δ: 2.07-2.12 (m, 1 H), 2.18-2.22 (m, 1 H), 2.70-2.76 (m, 1 H), 2.80-2.84 (m, 1 H), 3.03 (dd, *J* = 16.5, 4.5 Hz, 1 H), 3.48 (dd, *J* = 16.5, 5.5 Hz, 1 H), 4.48-4.51 (m, 1 H), 7.18-7.34 (m, 5 H); ^{13}C NMR (125 MHz, $CDCl_3$) δ: 31.1, 36.2, 42.7, 70.3, 126.2, 128.2, 128.5, 140.0, 168.0; MS (EI, 70 eV): *m/z* 176 (M^+), 158, 131, 117, 104, 91, 84; HRMS *m/z* calcd for $C_{11}H_{12}O_2$: 176.0838, found 176.0839; Anal. Calcd for $C_{11}H_{12}O_2$: C, 74.98; H, 6.86; Found: C, 74.97; H, 7.07.

24. Separation of the enantiomers by chiral HPLC (Daicel Chiracel™ OD-H column, flow rate 1.0 mL/min, 10% ^{i}PrOH, 90% hexane, T_r 14.3 (*S*) and 16.5 (*R*) min) provided the enantiomer ratio: 4(*S*):4(*R*) = 96.1:3.9 (92% ee).

Safety and Waste Disposal Information

All hazardous materials should be handled and disposed of in accordance with "Prudent Practices in the Laboratory"; National Academy Press; Washington, DC, 1995.

3. Discussion

Recent success in developing *de novo* asymmetric syntheses of enantioenriched β-lactones has created renewed interest in these heterocycles as versatile platforms for asymmetric organic synthesis.[3,4] β-Lactones are direct progenitors of numerous useful building blocks including enantioenriched β-amino acids,[5] allenes[6] and β,β-disubstituted carboxylic acids.[7] β-Lactones are also functional equivalents of ester aldol addition products.[8] In this latter context, we developed acyl halide-aldehyde cyclocondensation (AAC) reactions that deliver enantioenriched β-lactone acetate aldol surrogates from commercially available starting materials (eq 1).

We have recently described Al(III)-catalyzed cyclocondensations of acyl halides and aldehyde electrophiles as an operationally simple route to highly enantioenriched 4-substituted 2-oxetanones. The easily prepared Al(III)-triamine catalyst **3** is uniquely effective in mediating highly enantioselective [2+2] cycloadditions of *in situ* generated ketene and aldehydes. The enantioenriched triamine ligand **2** required for preparing the AAC catalyst is obtained in two high yielding steps from (*S*)-valinol. The AAC catalyst **3** is prepared from the triamine ligand *in situ* by reacting **2** with $AlMe_3$; the acetyl bromide, aldehyde and diisopropylethylamine required for the AAC reaction are then simply added to the resulting catalyst solution. The enantioenriched β-lactones emerging from the AAC reactions are typically sufficiently pure to be used in subsequent transformations without purification.

The catalyst complex **3** renders a variety of structurally diverse aldehydes as effective electrophiles for the catalyzed asymmetric AAC reactions. The procedure described herein highlights the reactivity of enolizable aliphatic aldehydes under the AAC reactions. The examples compiled in Table 1 are indicative of other aldehyde substrates that participate in efficient AAC reactions. Aromatic aldehydes bearing alkyl or electron-withdrawing substituents, functionalized aldehydes bearing common oxygen protecting groups and conjugated ynals are all very reactive electrophiles in the asymmetric AAC reactions. Aldehydes that are not useful AAC substrates include conjugated enals and α-branched aldehydes (c.f., cyclohexanecarboxaldehyde); these types of aldehydes afford little to no β-lactone product under the AAC reaction conditions.

Table 1

MeC(O)Br + HC(O)R → (10 mol% **3**, $^{i}Pr_2NEt$, CH_2Cl_2, - 50 °C) β-lactone

Aldehyde (R)	% yield	% ee
$-(CH_2)_8CH=CH_2$	91	91
$-CH_2CH(CH_3)_2$	80	93
$-CH_2OCH_2Ph$	91	92
$-CH_2OSiPh_2tBu$	74	89
$-CH_2CH_2OCH_2Ph$	90	91
$-C{\equiv}CCH_2OCH_2Ph$	86	93
$-C_6H_4NO_2$	93	95

1. Department of Chemistry, University of Pittsburgh, Pittsburgh, Pennsylvania, 15260.
2. Cernerud, M.; Skrinning, A.; Bérgère, I.; Moberg, C. *Tetrahedron: Asymm.* **1997**, *8*, 3437-3441.
3. Acyl halide-aldehyde cyclocondensations: (a) Nelson, S. G. Peelen, T. J.; Wan, Z. *J. Am. Chem. Soc.* **1999**, *121*, 9742. (b) Nelson, S. G.; Wan, Z. *Org. Lett.* **2000**, *2*, 1883. (c) Nelson, S. G.; Zhu, C.; Shen, X. *J. Am. Chem. Soc.* **2004**, *126*, 14-15.

4. Leading references to catalytic asymmetric β-lactone preparation: (a) Wynberg, H.; Staring, E. G. J. *J. Am. Chem. Soc.* **1982**, *104*, 166. (b) Romo, D.; Harrison, P. H. M.; Jenkins, S. I.; Riddoch, R. W.; Park, K.; Yang, H. W.; Zhao, C.; Wright, G. D. *Bioorg. Med. Chem.* **1998**, *6*, 1255. (c) Tennyson, R.; Romo, D. *J. Org. Chem.* **2000**, *65*, 7248. (d) Evans, D. A.; Janey, J. M. *Org. Lett.* **2001**, *3*, 2125. (e) Cortez, G. S.; Tennyson, R. L.; Romo, D. *J. Am. Chem. Soc.* **2001**, *123*, 7945. (f) Calter, M. A.; Liao, W. *J. Am. Chem. Soc.* **2002**, *124*, 13127 and references therein.
5. Nelson, S. G.; Spencer, K. L. Angew. *Chem. Int. Ed.* **2000**, 39, 1323.
6. Wan, Z.; Nelson, S. G. *J. Am. Chem. Soc.* **2000**, *122*, 10470.
7. Nelson, S. G.; Wan, Z.; Stan, M. A. *J. Org. Chem.* **2002**, *67*, 4680.
8. Nelson, S. G.; Wan, Z.; Peelen, T. J.; Spencer, K. L. *Tetrahedron Lett.* **1999**, *40*, 6535.

Appendix
Chemical Abstracts Nomenclature (Registry Number)

(*S*)-Valinol: (2*S*)-2-Amino-3-methyl-1-butanol; (2026-48-4)

Triethylamine: *N,N*-Diethylethanamine,; (121-44-8)

Trifluoromethanesulfonic anhydride; (358-23-6)

(*S*)-*N*-Trifluoromethylsulfonyl-2-isopropylaziridine: (*S*)-2-(1-Methylethyl)-1-[(trifluoromethyl)sulfonyl]aziridine; (196520-85-1)

Benzylamine: Benzenemethanamine; (100-46-9)

(*2S,6S*)-4-Benzyl-1,7-bis(trifluoromethylsulfonyl)-2,6-diisopropyl-1,4,7-triazaheptane: *N,N'*-[[(Phenylmethyl)imino]bis[(1*S*)-1-(1-methylethyl)-2,1-ethanediyl]]bis[1,1,1-trifluoro]-methanesulfonamide; (200351-80-0)

Diisopropylethylamine: *N*-Ethyl-*N*-(1-methylethyl)-2-propanamine; (7087-68-5)

Trimethylaluminum; (75-24-1)

Acetyl bromide; (506-96-7)

Hydrocinnamaldehyde: Benzenepropanal; (104-53-0)

(4*S*)-4-(2-Phenethyl)oxetan-2-one:, (4*S*)-4-(2-phenylethyl)-2-oxetanone; (214853-90-4)

ALKYNE *VIA* SOLID-LIQUID PHASE-TRANSFER CATALYZED DEHYDROHALOGENATION: ACETYLENE DICARBOXALDEHYDE TETRAMETHYL ACETAL AND ACETYLENE DICARBOXALDEHYDE DIMETHYL ACETAL

[1,1,4,4-Tetramethoxy-2-butyne and 4,4-Dimethoxy-2-butynal]

A. 1 → 2: Br_2, CH_2Cl_2; 10 °C, 90 min

B. 2 → 3: 1. H_2SO_4, CH_3OH, reflux, 72 hr; 2. $(CH_3CH_2)_3N$

C. 3 → 4: KOH, TMEEA; THF, reflux, 24 hr

D. 4 → 5: HCO_2H, CH_2Cl_2; 16-19 °C, 60 hr

Submitted by Rufine Akué-Gédu and Benoît Rigo.[1]
Checked by Shigeki Matsunaga and Masakatsu Shibasaki.

1. Procedure

A. *3,4-Dibromo-2,5-dimethoxytetrahydrofuran* (**2**). A 1-L, three-necked, round–bottomed flask, equipped with a magnetic stirring bar, thermometer, pressure-equalizing addition funnel and water condenser connected to nitrogen, is loaded with 150 mL of dichloromethane (Note 2) and 2,5-dimethoxy-2,5-dihydrofuran (**1**) (50 g, 0.384 mol) (Note 3). The flask is cooled to 10°C by using an ice bath. Bromine (61.4 g, 0.384 mol) is added over 120 min, while keeping the reaction temperature below 15°C. After addition is complete, the reaction mixture is stirred at 10°C until the bromine color disappears. The solution is concentrated on a rotary evaporator while keeping the bath temperature below 20°C (Note 4). 3,4-Dibromo-2,5-dimethoxytetrahydrofuran (**2**) is obtained as a light colored semi-solid mass (110-111 g, 99%, mixtures of isomers) (Notes 5 and 6).

B. *2,3-Dibromo-1,1,4,4-tetramethoxybutane* (**3**). A 3-L, three-necked, round–bottomed flask, equipped with a heating mantle, a pressure-equalizing addition funnel and a water condenser connected to nitrogen, is loaded with crude 3,4-dibromo-2,5-dimethoxytetrahydrofuran (**1**) prepared in Step A (110-111 g), 2 L of methanol and boiling chips. The solution is heated to reflux, at which time 96% sulfuric acid (37.7 g, 0.384 mol) is added over 30 min (Notes 7 and 8). The solution is refluxed for 72 hours, then cooled to room temperature. Triethylamine (49 mL, 0.352 mol) is added (Notes 9 and 10), then the solution is concentrated on a rotary evaporator. Heptane (250 mL) is added and the residue is stirred and refluxed for 20 min. The lightly colored heptane phase is separated using a separatory funnel. The heptane extraction step is repeated three times. The combined heptane solutions are concentrated on a rotary evaporator to give 2,3-dibromo-1,1,4,4-tetramethoxybutane (**3**) as a yellow oil (99-100 g including inpurities, <79%) (Note 11).

C. *1,1,4,4-Tetramethoxybut-2-yne* (**4**). A 1-L, one-necked, round-bottomed flask is equipped with a powerful magnetic stirrer, a large rugby ball-shaped stirring bar, and a water condenser equipped with a nitrogen inlet. The flask is charged with 300 mL of tetrahydrofuran (Note 3), tris[2-(2-methoxyethoxy)ethyl]amine (TMEEA, 10.1 g, 31.2 mmol) (Note 3) and crude 2,3-dibromo-1,1,4,4-tetramethoxybutane (**3**) prepared in Step B (99 g). Potassium hydroxide pellets (85%) (70 g, 1.06 mol) (Note 12) are then added to the homogenous solution, and the stirred mixture is refluxed for 24 hr using an oil-bath. Solvent is evaporated on a rotary evaporator, water (250 mL) is added, and the solution is partitioned three times with ether (250 mL) (Note 13). The combined ether phases are washed three times with water (100 mL). After the solution is dried over $MgSO_4$, the solution is concentrated on a rotary evaporator to give 1,1,4,4-tetramethoxybut-2-yne (**4**) as a yellow oil (50-51 g, 97-99%) (Notes 14, 15, and 16).

D. *4,4-Dimethoxybut-2-ynal* (**5**). A 1-L round-bottomed flask, charged with 150 mL of dichloromethane (Note 2) and crude 1,1,4,4-tetramethoxybut-2-yne (**4**) (50 g, 287 mmol) prepared in Step C is cooled to 0°C in an ice bath. Another flask charged with formic acid (280 g, 6 mol), 150 mL of dichloromethane, and 1.5 L of H_2O is also cooled to 0°C. The cooled solution of formic acid is then rapidly added to the diacetal solution. The neck is sealed with a rubber septum equipped with a stainless steel syringe needle (Note 17). The flask is rapidly covered by a large opaque towel, and the reaction is stirred in the dark at 16-19°C using a temperature regulated water bath for 60 hr (Note 18). The dark brown solution is partitioned three times with ice-cooled water (1 x 200 mL, then 2 x 150 mL). The aqueous phases are combined and extracted three times with dichloromethane (100 mL). The dichloromethane extracts are combined, washed with ice-cooled water (100 mL), and then dried with a mixture of Na_2SO_4 /$NaHCO_3$ (10/1). The solution is concentrated on a rotary evaporator while keeping the bath temperature below 25°C (Note 4), to give 4,4-dimethoxybut-2-ynal (**5**) (25.5 g, 69% from **4**, 52% from **1**) as a brown oil (Note 19, 20, 21, 22, 23).

2. Notes

1. This procedure is an adaptation of a procedure published earlier by our group.[2]

2. Dichloromethane was distilled from P_2O_5.

3. 2,5-Dimethoxy-2,5-dihydrofuran (99%) was purchased from Alfa Aesar; triethylamine (99.5%) was purchased from Aldrich; tetrahydrofuran (99+%) was purchased from Acros; TMEEA, (tris[2-(2-methoxyethoxy)ethyl]amine) (99%) was purchased from Alfa Aesar; formic acid (95-97%) was purchased from Aldrich. These compounds were used as received.

4. Heating the oil bath of the rotary evaporator leads to a more colored, less pure product.

5. While the product was sufficiently pure for the subsequent step, ^{1}H NMR indicated the presence of undetermined impurities and showed that a mixture of the three isomers of 3,4-dibromo-2,5-dimethoxytetrahydrofuran was obtained in a ratio depending mainly on the temperature and length of the evaporation of solvent (The ratio of products as determined by ^{1}H NMR 43/17/40 (submitters), 46/12/42 (checkers); Isomer 1 (46% of the mass, mp 88°C) ^{1}H NMR ($CDCl_3$) δ: 3.48 (s, 3 H), 3.50 (s, 3 H), 4.16 (dd, J = 9.6, 3.9 Hz, 1 H), 4.26 (dd, 9.6, 3.9 Hz, 1 H), 4.94 (d, J = 3.9 Hz, 1 H), 5.21 (d, J = 3.9 Hz, 1 H); Isomer 2 (12% of the mass, mp 56°C) ^{1}H NMR ($CDCl_3$) δ: 3.47 (s, 6 H), 4.42 (dd, J = 1.9, 0.7 Hz, 2 H), 5.23 (dd, 1.9, 0.7 Hz, 2 H); Isomer 3 ^{1}H NMR ($CDCl_3$) δ: (42% of the mass, mp 72°C) 3.50 (s, 6 H), 4.18 (dd, J = 1.9, 1.3 Hz, 2 H), 5.29 (dd, 1.9, 1.3 Hz, 2 H).

6. The submitter describes the following unchecked procedure to separate three isomers; the mixture of three isomers of **2** can be partially separated by cooling the crude semisolid mass at 0°C overnight. Filtration of the solid and washing with cold (0°C) heptane, gave 78 g of isomer with mp 88-89°C. Heptane was added to the residue, and the solution was cooled at 0°C. Filtration and washing of the solid with cold (0°C) heptane, gave isomer with mp 72-74°C. Chromatography of the residue on alumina can give the third isomer (mp 56-58°C).[3]

7. If a lesser amount of sulfuric acid is used, the reaction takes a longer time and leads to less pure product.

8. Slow addition of sulfuric acid to the hot reaction mixture allows for better control of the exothermicity of the addition, and provides a less colored final product.

9. Neutralization of sulfuric acid before evaporation of methanol avoids decomposition and colorization of diacetal **3**.

10. Triethylamine was selected for the neutralization because it decomposes the methyl sulfate formed from reaction of sulfuric acid with methanol.

11. Product **3** is difficult to purify, but remaining impurities are removed during the treatment of the next step. ^{1}H NMR ($CDCl_3$) δ: 3.44 (two singlet peaks overlapped with impurities, 12 H), 4.34 (d, J = 7.9 Hz, 2 H), 4.58 (d, J = 7.9 Hz, 2 H).

12. Potassium hydroxide pellets are better than finely ground powder, because the powder readily solidifies into a mass causing the reaction to slow down.

13. It was necessary to use diethyl ether instead of CH_2Cl_2 to avoid extraction of the phase transfer agent.

14. It was possible to perform the dehydrobromination step with potassium *tert*-butoxide in DMSO (2 hr, 20°C in a smaller scale), but treatment of the reaction mixture was less efficient and the purity of the product was lower.

15. While the product was sufficiently pure for the subsequent step, ^{1}H NMR indicated the presence of a small amount of undetermined impurities. If very pure product is needed for another use, compound **4** can be purified by distillation through a 10-cm Vigreux column: bp 55-59°C/0.5 mm [bp = 70-73°C/1 mm,[4] bp = 101-103°C/13 mm[5]] to give a colorless liquid. ^{1}H NMR ($CDCl_3$) δ: 3.40 (s, 12 H), 5.21 (s, 2 H).

16. Alternatively, it was not necessary to perform heptane extraction of compound **3**: the solution obtained after neutralization **3** with

triethylamine (Step B) was evaporated on a rotary evaporator. Tetrahydrofuran (250 mL) (Note 2), TMEEA (0.03 mol, 10.1 g) (Note 2) and pellets of 85% potassium hydroxide (110 g, 1065 mol) (Note 11) were added. The mixture was refluxed for 24 hours, and then part of THF was evaporated. Water was added and the mixture was extracted with ether, leading to 56 g of **4**. The weight of diacetal **4,** including impurities, is higher with no heptane extraction (56 g for Steps B and C); however, the purity of **4** is higher using the heptane extraction procedure.

17. The syringe needle is used for carbon dioxide evolution from the reaction mixture containing formic acid and acetal **4**.

18. It is important to perform the deprotection step below 20°C in the dark, because exposure of the reaction mixture to a higher temperature or to light can lead to a lower yield through formation of a highly colored by-product. The reaction time and yield of Step D strongly depended on the quality of formic acid used. 95-97% formic acid from Aldrich gave best results. Drier formic acids: 99% from Acros Organics (Cat. N° 14793), 98% (Cat. N° 8841) from Lancaster, 97% from Avocado (Cat. N° 13285), gave decomposition of the monoacetal and formation of black tars. Even using 95-97% formic acid from Aldrich, difference in lot number may cause different yield and reaction time. Although addition of a small amount of H_2O (1.5 mL) was not a part of the submitter's original procedure, the isolated yield of **5** by the checkers is slightly improved by addition of a small amount of H_2O to the reaction mixture. Loss of product during work-up can also be problematic. It is necessary to wash the organic extracts with ice-cooled water quickly to avoid decomposition of product during extraction and washing. The temperature of the water bath also affected the reaction time significantly. The reaction was judged to be complete after 60 h at 16-19°C (by the checkers), after 32 h at 17-20°C (by the checkers), and after 20 h at 20-21°C (by the submitters).

19. Because the formic acid quality, the reaction temperature, the quality of crude **4**, and the amount of water greatly influence the reaction time and isolated yield, it is essential to follow the reaction by NMR analysis of the crude reaction mixture. A 0.5 mL sample of the reaction mixture is evaporated under reduced pressure (2 mm) below 20°C and dissolved in $CDCl_3$. A resonance at 5.20 ppm ($CH(OMe)_2$ of dialdehyde) is compared with resonances at 5.31 ppm {$CH(OMe)_2$} and 9.27 ppm {(CHO) of the monoaldehyde}

20. ^1H NMR spectrum of 4,4-dimethoxybut-2-ynal (**5**) indicated the presence of tiny amount of undetermined impurities. ^1H NMR ($CDCl_3$) δ:

3.41 (s, 6 H), 5.31 (s, 1 H), 9.27 (s, 1 H); ^{13}C NMR ($CDCl_3$) δ: 53.0, 83.0, 88.3, 92.7, 176.0. Purity of this crude compound is sufficient for participation in Michael reactions with nucleophiles. Aldehyde **5** remains unaltered when kept for years at -40°C.

21. While the brown color of the crude compound can be removed by filtration on silica, most impurities remain in the product. [A solution of the product in CH_2Cl_2 was applied to the top of a short column of flash-grade silica gel (200 g, 10-cm diam x 4 cm) loaded with CH_2Cl_2. The brown color remains on the top of the column upon washing with CH_2Cl_2. Last traces of solvent were removed at 0.5 mm to afford **5** as a slighly yellow liquid.]

22. If one needs a purer product, compound **5** can be purified by distillation: 24 g of aldehyde **5** and a rugby ball-shaped stirring bar were introduced in a 50-mL one-necked flask immersed in an oil-bath, and fitted with a 10-cm Vigreux column, a water condenser cooled at –10 to –20°C by a cooling regulator, and a receiving flask cooled at –40°C by cooling bath. Aldehyde **5** was distilled under vacuum (0.2 mm), while the temperature of the oil-bath was slowly raised from 25°C to 60°C by using a temperature regulator. Aldehyde **5**, (17.3 g (72 %), bp 34-37°C (0.2 mm) [lit bp = 75-78°C (14 mm)],[4]) was obtained as a slightly yellow liquid.

23. While working in a 20 to 50 g scale of diacetal **4**, compound **5** was obtained in 85-92% yields (by the submitters) and in 72% (by the checkers in 25-g scale).

Safety and Waste Disposal Information

All hazardous materials should be handled and disposed of in accordance with "Prudent Practices in the Laboratory"; National Academy Press; Washington, DC, 1995.

3. Discussion

This procedure describes the synthesis of acetylene dicarboxaldehyde mono- and bis-*dimethyl* acetals. It was reported[4,6] that acetylene dicarboxaldehyde *dimethyl* acetals are less reactive than their *diethyl* analogs. The new procedure takes advantages of this slightly lower reactivity and higher stability of the *dimethyl* acetals (Reaction time for the formic acid hydrolysis described in Step D is 20-60 hours when using *dimethyl* acetals, whereas it is 10 hours when using *diethyl* acetals). Acetylene dicarboxaldehyde[4-15] and its precursors mono and bis *diethyl* acetals are useful starting materials, which have found many applications in dipolar cycloaddition,[7-11] Diels-Alder,[4,6,7,12,13] Michael,[4,6,7,13] or Wittig[15] reactions.

A major drawback for a larger use of these compounds is their high price[16] or difficulties in their preparation, which preclude a large-scale synthesis. Two main procedures for the synthesis of these compounds are described in the literature. In a first reaction sequence, addition of bromine to acrolein followed by a dehydrobromination step leads to acetylene carboxaldehyde diethyl acetal. Reaction of this compound with triethyl orthoformate using $ZnCl_2$ afforded the ethyl analog of acetal **4**.[2] In order to obtain 50 g of diacetal using this method,[2,6,17] it would be necessary to use 175 g of triethyl orthoformate, 345 g of KOH and 575 g of tetrabutylammonium hydrogen sulfate. In another approach, acetylene was reacted with ethyl magnesium bromide to give a bis-magnesium salt. Reaction of this salt with triethyl orthoformate yields the ethyl analog of acetal **4**. In order to obtain 50 g of the diacetal using this second method,[4,6] it would be necessary to use 200 g of triethyl orthoformate, 19 g of magnesium and 260 L of gaseous acetylene. During this last procedure, it is necessary to avoid precipitation of the intermediate bis-magnesium salt of acetylene, which due to difficulties in stirring can lead to poor yields.[4]

The use of 99% formic acid to perform hydrolysis of one functional group of acetylene dicarboxaldehyde tetramethyl acetal is an adaptation of a method from Gorgues.[4,5,18]

The procedure described above reports an easy, high yielding, large-scale preparation of dimethyl acetals **4** from inexpensive 2,5-dimethoxy-2,5-dihydrofuran (**1**), a compound that contains all the carbon atoms of acetylene dicarboxaldehyde. The use of the phase transfer agent TMEEA is very important for this reaction,[19] because TMEEA enables the use of KOH and THF instead of DMSO and potassium *tert*-butoxide in the dehydrobromination step.

1. Groupe de Recherche sur l'Inhibition de la Prolifération Cellulaire (EA 2692), Ecole des Hautes Etudes Industrielles, 13 rue de Toul, 59046 Lille, France.

2. Akué-Gédu, R.; Rigo B. *Tetrahedron Lett.,* **2004**, *45*, 1829.
3. Gagnaire, D.; Vottero, P. *Bull. Soc. Chim. Fr.* **1963**, 2779.
4. Gorgues, A.; Stephan, D.; Belyasmine, A.; Khanous, A.; Le Coq, A. *Tetrahedron* **1990,** *46*, 2817.
5. Wohl, A.; Bernreuter, E. *Ann. Chem.* **1930**, *481*, 1.

6. Gorgues, A.; Simon, A.; Le Coq, A.; Hercouet, A.; Corre, F. *Tetrahedron* **1986,** *42*, 351.
7. Gorgues, A. *Janssen Chim. Acta*, **1986,** *4*, 21.
8. Gorgues, A.; Le Coq, A. *Tetrahedron Lett.* **1979,** *20*, 4829.
9. Henkel, K.; Weygand, F. *Chem. Ber.* **1943**, *76B*, 812.
10. Ramanaiah, K.C.V.; Stevens, E.D.; Trudell, M.L.; Pagoria, P.F. *J. Heterocyclic Chem.* **2000**, *37*, 1597.
11. Frère, P.; Belyasmine, A.; Gouriou, Y.; Jubault, M.; Gorgues, A.; Duguay, G.; Wood, S.; Reynolds, C.D.; Bryce, M.R. *Bull. Soc. Chim. Fr.* **1995**, *132*, 975.
12. Gorgues, A.; Le Coq, A. *Chem. Commun.* **1979**, 767.
13. Stephan, D.; Gorgues, A.; Belyasmine, A.; Le Coq, A. *Chem. Commun.* **1988,** 263.
14. Blanco, L.; Bloch, R.; Bugnet, E.; Deloisy, S. *Tetrahedron Lett.* **2000,** *41*, 7875.
15. Barrett, A.G.M.; Hamprecht, D.; Ohkubo, M. *J. Org. Chem.* **1997**, *62*, 9376.
16. 3,3-Diethoxy-1-propyne: 329 €/25 g (Lancaster); acetylenedicarboxaldehyde tetraethyldiacetals: 75 €/1 g (Acros); acetylenedicarboxaldehyde monodiethylacetal: 78 €/1 g (Acros); 2,5-dimethoxy-2,5-dihydrofuran (**1**): 191 €/1000 g (Alfa Aesar).
17. (a) Le Coq, A.; Gorgues, A. *Org. Synth. Coll. Vol. VI*, **1988**, 954; (b) Montaña, A.M.; Fernandez, D.; Pagès, R.; Filippou, A.C.; Kociok-Köhn, G. *Tetrahedron* **2000**, *56*, 425.
18. Gorgues, A. *Bull. Soc. Chim. Fr.* **1974**, 529.
19. (a) Soula, G. *J. Org. Chem.* **1985**, *50*, 3717; (b) Stafford, A.; McMurry, J. E. *Tetrahedron Lett.,* **1988**, *29*, 2531. This phase transfer agent strongly accelerates dehydrobromination of alkyl halides in the presence of KOH. (c) Tamao, K.; Nakagawa, Y.; Ito, Y. *Org. Synth. Coll. Vol. IX* **1998**, 539.

Appendix
Chemical Abstracts Nomenclature; (Registry Number)

2,5-Dimethoxy-2,5-dihydrofuran: Furan, 2,5-dihydro-2,5-dimethoxy-; (332-77-4)

Bromine; (7726-95-6)

Dibromo-2,5-dimethoxytetrahydrofuran: Furan, 3,4-dibromotetrahydro-2,5-

dimethoxy-; (91468-55-2)

Methanol; (67-56-1)

Triethylamine: Ethanamine, *N,N*-diethyl-: (121-44-8)

2,3-Dibromo-1,1,4,4-tetramethoxybutane: Butane, 2,3-dibromo-1,1,4,4-tetramethoxy-; (25537-21-7)

Tris[2-(2-methoxyethoxy)ethyl]amine: Ethanamine, 2-(2-methoxyethoxy)-*N,N*-bis[2-(2-methoxyethoxy)ethyl]-; (70384-51-9)

Potassium hydroxide: (1310-58-3)

1,1,4,4-tetramethoxybut-2-yne: 2-Butyne, 1,1,4,4-tetramethoxy-; (53281-53-1)

Formic acid; (64-18-6)

4,4-Dimethoxybut-2-ynal: 2-Butynal, 4,4-dimethoxy-; (124744-10-1)

CATALYTIC REDUCTION OF AMIDES TO AMINES WITH HYDROSILANES USING A TRIRUTHENIUM CARBONYL CLUSTER AS THE CATALYST

(*N*,*N*-Dimethyl-3-phenylpropylamine)

A. acenaphthylene + $Ru_3(CO)_{12}$ → (heptane, reflux, 36 h) → **1**

B. **1** (1 mol% to **2**) in THP → ($PhMe_2SiH$ (2.2 eq. to **2**), rt, 30 min) → (**2**, $PhCH_2CH_2CONMe_2$; rt, 75 min) → **3** ($PhCH_2CH_2CH_2NMe_2$)

Submitted by Yukihiro Motoyama, Chikara Itonaga, Toshiki Ishida, Mikihiro Takasaki, and Hideo Nagashima.[1]
Checked by Joshua G. Pierce and Peter Wipf.

1. Procedure

Caution! Since reaction intermediates are potentially unstable to air and moisture, preparation of the ruthenium cluster should be carried out under an inert gas atmosphere.

A. *(μ_3, η^2: η^3: η^5-Acenaphthylene)$Ru_3(CO)_7$* (**1**). A two-necked, 200-mL round-bottomed flask is equipped with a magnetic stir bar, a reflux condenser, and a stopcock. The top of the condenser is fitted with a three-way stopcock, of which one way is connected to an argon flow line and the other is connected to a vacuum line. The apparatus is flame-dried while under vacuum, and allowed to cool to room temperature under an argon purge. The stopcock is removed, and the flask is charged with $Ru_3(CO)_{12}$ (Note 1) (639 mg, 1 mmol), acenaphthylene (Note 2) (182 mg, 1.2 mmol), and 75 mL of heptane (Note 3) under an argon stream. The stopcock is again placed on the flask, and the mixture is heated under reflux. At the initial stage, the solution contains a yellow supernatant and an undissolved orange solid of $Ru_3(CO)_{12}$. The color of the solution gradually turns to red and all of the solid materials are dissolved after 30 min (Note 4).

After 36 h, a dark red solution and some red precipitates have formed. The mixture is cooled to room temperature under an argon atmosphere. From this stage, subsequent work can be performed in the air. The precipitates are collected on a glass filter, washed with *n*-hexane (30 mL), and dissolved in dichloromethane (20 mL) (Note 5). Celite® (*ca.* 3 g) (Note 6) is added to this solution and stirred for 1 min, then the solvent is removed under reduced pressure. The crude ruthenium complex adsorbed to Celite® is applied to a pre-packed silica gel column (20 mm x 100 mm) (Note 7), then sea sand (10 mm) (Note 8) is loaded onto the top of the Celite® for preventing the powder from diffusing into the eluent. The silica gel is first washed with 80 mL of *n*-hexane to remove a small amount of unreacted acenaphthylene (yellow band), then the product (red-brown band) is eluted with 2.5 L of 2:1 ether:*n*-hexane (Notes 9 and 10) to afford 620 mg (95%) of (μ_3, η^2: η^3: η^5-acenaphthylene)$Ru_3(CO)_7$ (**1**) (Note 11), which can be used for reduction of amides without further purification. Dark red crystals of **1** are obtained by recrystallization of the product obtained as above from a mixture of dichloromethane and *n*-hexane at -30°C (Note 12).

*Caution! The cluster **1** is first treated with excess hydrosilane for activation. Treatment of the activated catalyst with amides in the presence of excess hydrosilane is exothermic. The reaction apparatus should be equipped with an efficient reflux condenser, and the amide should be added carefully to avoid overheating of the reaction mixture.*

B. *N,N-Dimethyl-3-phenylpropylamine* (**3**). A 200-mL, two-necked round-bottomed flask is equipped with an efficient reflux condenser and a septum. The top of the reflux condenser is fitted with a three-way stopcock, of which one way is connected to an argon flow line and the other is connected to a vacuum line. The apparatus is flame-dried *in vacuo*, and allowed to cool to room temperature under an argon purge. The septum is removed, and the flask is charged with **1** (365 mg, 0.557 mmol, 1 mol% based on *N,N*-dimethyl-3-phenylpropionamide (**2**) (Note 13)) under an argon flow. The septum is again placed on the flask, and tetrahydropyran (THP) (4.5 mL) (Notes 14 and 15) is added to dissolve the catalyst. Phenyldimethylsilane (19 mL, 123 mmol) (Note 16) is charged *via* syringe through the septum, and the mixture is stirred for 30 min at room temperature. At this point, the initial dark-orange color of the catalyst solution changed to light orange. *N,N*-Dimethyl-3-phenylpropionamide (**2**) (10 mL, 56 mmol) (Note 17) is added drop-wise over a period of 75 min *via* syringe (Note 18). *Caution! The reaction is exothermic* (Note 19). The solution is stirred for an additional 75 min, resulting in consumption of the amide as determined by TLC analysis (Note 20). The solvent is removed by rotary evaporator at room temperature (70 Torr) and the residue is poured into a 500-mL Erlenmeyer flask containing conc. HCl (84 mL, 12 M) at 0°C. The reaction flask is rinsed with ether (3 x 84 mL), and the liquid is added to the Erlenmeyer flask. The mixture is moved to a separatory funnel and shaken well. The resulting aqueous solution is washed three times with 84 mL of ether (total of 252 mL). All ether layers are discarded and the aqueous phase is moved to a 500-mL Erlenmeyer flask, cooled in an ice bath, and basified with KOH pellets (84 g) with stirring to form white solids. Ether (164 mL) is added to the mixture, stirred for 20 min, and the organic layer is separated. This ether extraction is repeated four times (84 mL each) (Note 21). The combined organic phases are dried over anhydrous Na_2SO_4 and concentrated under reduced pressure at room temperature (50 Torr). Distillation of the crude product at 57°C (2.1 Torr) gives *N,N*-dimethyl-3-phenylpropylamine (**3**) as a colorless oil (Notes 22 and 23) (8.32 g, 91%).

2. Notes

1. $Ru_3(CO)_{12}$ (99%) is commercially available from Aldrich Chemical Company, Inc., although the submitters have prepared it according to a reported procedure.[2]

2. Acenaphthylene was purchased from Aldrich Chemical Company, Inc. (99+%) or Tokyo Kasei Kogyo Company, Ltd. (99+%) and used as received.

3. Heptane (99+%) was purchased from Kanto Chemical Company, Inc. and distilled from calcium hydride before use. Checkers purchased heptane from Aldrich Chemical Company.

4. The progress of the reaction was followed by TLC analysis (Merck Art No. 5715: Silica gel 60 F_{254}, 0.25 mm thickness) with 2:1 ether:*n*-hexane as eluent and visualization with a UV lamp at 254 nm. The acenaphthylene and $Ru_3(CO)_{12}$ starting materials have an R_f = 0.89 and 0.95, respectively, and the complex **1** has an R_f = 0.42.

5. *n*-Hexane (96%) and dichloromethane (99%) were purchased from Kishida Chemical Co., Ltd. (submitters) or VWR (checkers), and used as received.

6. Celite 545 was purchased from Kishida Chemical Company, Ltd. (submitters) or Fisher Scientific (checkers), and used as received.

7. Silica gel 60 (70-230 mesh) (Merck Art No. 7734) was purchased

from Merck (submitters) or Sorbent Technologies (checkers), and used as received.

8. Sea sand (425-850 μm; 20-35 mesh) was purchased from Wako Pure Chemical Industries, Ltd. (submitters) or Aldrich (checkers), and used as received.

9. Ether (99.5%) was purchased from Kishida Chemical Co., Ltd. (submitters) or VWR (checkers), and used as received.

10. Although the submitters prefer ether to dichloromethane for environmental reasons, a 1:1 mixture of dichloromethane:*n*-hexane (600 mL) can alternatively be used as the eluent for collection of a second red-brown band. Use of dichloromethane instead of ether can reduce the total amounts of eluent. TLC analysis (Merck Art No. 5715 : Silica gel 60 F_{254}, 0.25 mm thickness) was performed with 50% dichloromethane-hexane as eluent and visualization with UV lamp at 254 nm. The acenaphthylene and $Ru_3(CO)_{12}$ starting materials have $R_f = 0.86$ and 0.91, and the complex **1** has $R_f = 0.22$.

11. The product exhibits the following physical and spectral properties: mp 231.2-232.1°C (dec). ^{1}H NMR (500 MHz, $CDCl_3$) δ: 2.52 (d, 1 H, $J = 6.1$ Hz), 4.90 (d, 1 H, $J = 2.5$ Hz), 5.48 (d, 1 H, $J = 6.0$ Hz), 5.58 (t, 1 H, $J = 6.3$ Hz), 5.81 (d, 1 H, $J = 9.2$ Hz), 6.01 (d, 1 H, $J = 2.5$ Hz), 6.25 (dd, 1 H, $J = 9.2$, 6.1 Hz), 6.48 (d, 1 H, $J = 6.3$ Hz). ^{13}C NMR (76 MHz, THF-d_8, -60 °C) δ: 37.7, 66.1, 67.0, 69.0, 78.5, 80.1, 82.2, 84.8, 88.3, 99.1, 116.1, 130.4, 188.7 (CO), 192.2 (CO), 195.7 (CO), 204.0 (CO), 204.7 (CO), 207.4 (CO), 265.9 (μ-CO. IR (KBr) cm^{-1} ν_{CO} 2039 (s), 2028 (s), 2012 (s), 1996 (s), 1986 (s), 1946 (s), 1768 (m: μ-CO). The purity (>98%) was determined by ^{1}H NMR and mp.

12. The complex (624 mg) is dissolved in 30 mL of dichloromethane at 40°C, then cooled at room temperature. To this solution, 70 mL of *n*-hexane is carefully added and the two-phase solution is stored at –37°C overnight. Orange crystals are collected by suction filtration on a Büchner funnel, washed three times with 20 mL of *n*-hexane (total of 60 mL), and then transferred to a 20-mL, round-bottomed flask and dried for 4 h at 0.01 Torr to provide 497 mg (80%) of **1** as orange microcrystals, which can be stored in the air for years without decomposition.

13. At lower catalyst concentration, formation of an unidentified byproduct decreases the yield of the product. For example, use of 0.1 mol% of the catalyst to **2** affords the desired amine in about 60% yield.

14. Tetrahydropyran (THP) (98+%) was purchased from Kanto Chemical Company, Inc. (submitters) or Aldrich (checkers), and distilled

from benzophenone ketyl before use. Anhydrous toluene, diethyl ether, and 1,4-dioxane (a cancer suspected reagent) can alternatively be used as the solvent.

15. The ruthenium cluster activated by hydrosilane is an efficient catalyst for conversion of hydrosilanes to silanols and siloxanes. The reduction should be performed in anhydrous THP under an inert gas atmosphere to avoid contact with moisture.

16. Phenyldimethylsilane (98+%) was purchased from Aldrich Chemical Company, Inc. and distilled before use.

17. *N,N*-Dimethyl-3-phenylpropionamide is commercially available in milligram quantities (TimTec Overseas Stock, Ambinter Screening Library, MicroChemistry Screening Collection), but it can prepared from dihydrocinnamoyl chloride (95+%) (purchased from Tokyo Kasei Kogyo Company, Ltd. (submitters) or Aldrich (checkers)) and dimethylamine (99+%) (purchased from Aldrich Chemical Company, Inc.) by application of a standard method and distilled before use (bp 108°C / 2 Torr). The purity (>99%) was determined by ^{1}H NMR, and the submitters also used capillary GLC with a TC-WAX column (0.25 mm x 30 m) (oven temperature: 230°C; head pressure: 60 kPa; retention time: 11.8 min).

18. Although a syringe was used for the addition of amide **2**, a syringe pump or an addition funnel can also be used.

19. Although the submitters controlled the exothermic reaction by the addition rate of the amide, a water bath can also be used to cool the reaction mixture.

20. The progress of the reaction was followed by TLC analysis (Merck Art No. 5715: Silica gel 60 F_{254}, 0.25 mm thickness) with ethyl acetate as eluent and visualization with UV lamp at 254 nm. The amide starting material has an $R_f = 0.48$ and the amine product has an $R_f = 0.19$.

21. The submitters reported the reduction gave a single product (determined by ^{1}H NMR) in an earlier paper,[3] but isolated yields were somewhat lower (75%). This can be attributed to a loss of the product during the chromatographic removal of siloxane byproducts. In the improved procedure reported herein, separation of the silane products from the desired amine is achieved by simple extraction, which contributes to an improvement in the yield of the product.

22. The product **3**[4] exhibits the following physical and spectral properties: ^{1}H NMR (500 MHz, $CDCl_3$) δ: 1.45-1.78 (m, 2 H), 2.24 (s, 6 H), 2.31 (t, 2 H, J = 7.3 Hz), 2.65 (t, 2 H, J = 7.7 Hz), 7.18-7.22 (m, 3 H), 7.27-7.32 (m, 2 H); ^{13}C NMR (76 MHz, $CDCl_3$) δ: 29.4, 33.5, 45.4, 59.2,

125.6, 128.2, 128.3, 142.2. IR (neat) cm^{-1} 3062, 3026, 2942, 2764, 1603, 1496, 1454, 1265, 1030. EIMS *m/z* 163 (M^+, 40), 91 (60), 84 (65), 74 (60), 59 (100), 58 (90); HRMS (EI) *m/z* calcd for $C_{11}H_{17}N$ 163.1376, found 163.1361. The purity (>99%) was determined by ^1H NMR, and the submitters also used capillary GLC with a TC-WAX column (0.25 mm x 30 m) (oven temperature: 120 °C; head pressure: 60 kPa; retention time: 16.1 min).

23. The submitters measured the amount of residual ruthenium species in the product by ICP-MS to be less than *ca.* 3 ppm.

Safety and Waste Disposal Information

All hazardous materials should be handled and disposed of in accordance with "Prudent Practices in the Laboratory"; National Academy Press; Washington, DC, 1995.

3. Discussion

The reduction of amides has been generally accomplished by aluminum hydrides such as $LiAlH_4$ in THF or Et_2O, DIBAL-H, and Red-Al in benzene at room temperature.[5,6] Alanes, AlH_3 in Et_2O and $AlH_3 \cdot Et_3N$, or boranes, $BH_3 \cdot THF$ and $BH_3 \cdot SMe_2$ are also useful.[6] However, these reagents are air- and moisture sensitive, and careful handling is required. It is also well known that separation of the product from aluminum or boron wastes is often troublesome. Alkali borohydrides are rather stable and their handling is easy; however, reduction of amides with $NaBH_4$ requires application of high temperatures or addition of activators. For example, reduction of amides is achieved by $NaBH_4$ in the presence of Lewis acids such as $TiCl_4$ and $ZrCl_4$.[5] Activation of amides is an alternative method for successful borohydride reduction; a two step procedure consisting of iminium salt formation from amides by treatment with Et_3OBF_4 or $POCl_3$ followed by reduction with $NaBH_4$ is reported.[7]

Hydrosilanes are stable reducing agents and their handling is easy. Although a number of catalytic hydrosilylations of ketones and aldehydes have been reported, in which addition of the H-Si bond to the C=O moiety is catalyzed by transition metal salts or complexes, there have been few reports for the reduction of amides. The $RhCl(PPh_3)_3$-catalyzed reduction of amides has recently been reported by Ito and coworkers; the reaction requires the use of dihydrosilanes or trihydrosilanes.[8] Removal of the silicon waste from the reaction mixture is not easy. Fuchikami reported the reduction of amides with trialkylsilanes, which are rather convenient to manipulate compared to dihydrosilanes or trihydrosilanes, in the presence of several ruthenium catalysts; however, the reaction proceeds under drastic conditions.[9] Thus, the procedure using trialkylsilanes, which are stable and easy to work with, deserves further attention, especially when the reduction of amides can be achieved under mild conditions.

The ruthenium complex **1**[10] is a useful catalyst for hydrosilylation of ketones and aldehydes,[10] reduction of carboxylic acids, esters, and amides with hydrosilanes,[11] and reduction of acetals and cyclic ethers with hydrosilanes.[3] It is also a good catalyst for silane-induced ring-opening polymerization of cyclic ethers,[3] cyclic siloxanes,[12] and vinyl ethers.[13] Treatment of **1** with molecular hydrogen followed by treatment with carbon monoxide provides 4,5-dihydroacenaphthylene, which has never been synthesized by other methods.[14]

The amide reduction is accomplished under mild conditions by the use of **1** as the catalyst. Phenyldimethylsilane as the reducing reagent has the advantage of stability towards air and moisture, and it is easy to handle. It is known that the siloxane byproducts are sometimes difficult to separate in many silane-mediated reductions of carbonyl compounds. In the present procedure, the siloxane byproducts are easily removed by extraction. Thus, the present method is an easy way to synthesize amines from amides.

1. Institute for Materials Chemistry and Engineering, Graduate School of Engineering Sciences, Kyushu University, Kasuga, Fukuoka 816-8580, Japan.
2. Bruce, M. I.; Jensen, C. M.; Jones, N. L. *Inorg. Synth.* **1990**, *28* (Reagents Transition Met. Complex Organomet. Synth.), 216-220, and references cited therein.
3. Matsubara, K.; Iura, T.; Maki, T.; Nagashima, H. *J. Org. Chem.* **2002**, *67*, 4985.
4. Kuehne, M. E.; Shannon, P. J. *J. Org. Chem.* **1977**, *42*, 2082.
5. Cope, A. C.; Ciganek, E. *Org. Synth. Coll. Vol.* **1963**, *4*, 339.
6. Seyden-Penne, J. *Reductions by the Alumino- and Borohydrides in Organic Synthesis*; Wiley-VCH (1997).
7. Borch, R. F. *Tetrahedron Lett.* **1968**, 61.
8. Kuwano, R.; Takahashi, M.; Ito, Y. *Tetrahedron Lett.* **1998**, *39*, 1017.
9. Igarashi, M.; Fuchikami, T. *Tetrahedron Lett.* **2001**, *42*, 1945.
10. Nagashima, H.; Fukahori, T.; Aoki, K.; Itoh, K. *J. Am. Chem. Soc.* **1993**,

115, 10430.

11. Nagashima, H.; Suzuki, A.; Iura, T.; Ryu, K.; Matsubara, K. *Organometallics* **2000**, *19*, 3579.
12. Matsubara, K.; Terasawa, J.; Nagashima, H. *J. Organomet. Chem.* **2002**, *660*, 145.
13. Nagashima, H.; Itonaga, C.; Yasuhara, J.; Motoyama, Y.; Matsubara, K. *Organometallics* **2004**, *23*, 5779.
14. Nagashima, H.; Suzuki, A.; Nobata, K.; Itoh, K. *J. Am. Chem. Soc.* **1996**, *118*, 687.

Appendix
Chemical Abstract Nomenclature; (Registry Number)

Acenaphthylene; (208-96-8)

$Ru_3(CO)_{12}$: Ruthenium, dodecacarbonyltri-, triangulo; (15243-33-1)

(μ_3, η^2: η^3: η^5-acenaphthylene)$Ru_3(CO)_7$; Ruthenium,

[μ3-[(1,2,2a,8a,8b-η:3,4-η:5,5a,6-η)-acenaphthylene]]-μ-carbonyl-

hexacarbonyltri-, triangulo; (151364-75-9)

Phenyldimethylsilane: Silane, dimethylphenyl-; (766-77-8)

N,N-Dimethyl-3-phenylpropionamide: Benzenepropanamide,

N,N-dimethyl-; (5830-31-9)

N,N-Dimethyl-3-phenylpropylamine: Benzenepropanamine, *N,N*-dimethyl-;

(1199-99-1)

CUMULATIVE AUTHOR INDEX FOR VOLUMES 80-82

This index comprises the names of contributors to Volumes **80**, **81,** and **82**. For authors of previous volumes, see either indices in Collective Volumes I through X, or the single volume entitled *Organic Syntheses, Collective Volumes I-VIII, Cumulative Indices,* edited by J. P. Freeman.

CUMULATIVE SUBJECT INDEX FOR VOLUMES 80-82

This index comprises subject matter for Volumes **80**, **81**, and **82.** For subjects in previous volumes, see either the indices in Collective Volumes I through X or the single volume entitled *Organic Syntheses, Collective Volumes I-VIII, Cumulative Indices,* edited by J. P. Freeman.

The index lists the names of compounds in two forms. The first is the name used commonly in procedures. The second is the systematic name according to **Chemical Abstracts** nomenclature, accompanied by its registry number in parentheses. Also included are general terms for classes of compounds, types of reactions, special apparatus, and unfamiliar methods.

Most chemicals used in the procedure will appear in the index as written in the text. There generally will be entries for all starting materials, reagents, intermediates, important by-products, and final products. Entries in capital letters indicate compounds, reactions, or methods appearing in the title of the preparation.

Acenaphthylene; (208-96-8) **82**, 188

(μ_3, η^2: η^3: η^5-acenaphthylene)$Ru_3(CO)_7$; Ruthenium, [μ3-[(1,2,2a,8a,8b-η:3,4-η:5,5a,6-η)-acenaphthylene]]-μ-carbonyl-hexacarbonyltri-, triangulo; (151364-75-9) **82**, 188

Acetamido-TEMPO: 1-Piperidinyloxy, 4-(acetylamino)-2,2,6,6-tetramethyl-; (14691-89-5) **82**, 80

ACETIC ACID 2-METHYLENE-3-PHENETHYLBUT-3ENYL ESTER:: Benzenepentanol, β,γ-bis(methylene)-, acetate; (445234-76-4), **81**, 1

Acetic acid 5-phenylpent-2-ynyl ester: 2-Pentyn-1-ol, 5-phenyl-, acetate; (445234-71-9), **81**, 3

Acetic anhydride: Acetic acid, anhydride; (108-24-7), **80**, 177; **81**, 3; **82**, 115

2-Acetoxy-1,4-dioxane: 1,4-Dioxan-2-ol, acetate; (1743-23-3) **82**, 99

α-ACETOXY ETHER SYNTHESIS, **80**, 177

4-Acetylamino-2,2,6,6-tetramethylpiperidine-1-oxoammonium perchlorate: Piperidinium, 4-(acetylamino)-2,2,6,6-tetramethyl-1-oxo-, perchlorate (9); (219543-08-5) **82**, 80

4-Acetylamino-2,2,6,6-tetramethylpiperidine-1-oxoammonium tetrafluoroborate (219543-09-6) **82**, 80

Asymmetric hydrogenation, **81**, 178
Aziridine Formation; **82**, 170

Benzaldehyde; (100-52-7), **80**, 160
Benzamidine hydrochloride: Benzenecarboximidamide, monohydrochloride (9); (1670-14-0), **81**, 105
Benzenesulfenyl chloride; (931-59-9), **81**, 244, 246
Benzenethiol; Thiophenol; (108-98-5), **80**, 184
(4*S*)-2-(BENZHYDRYLIDENAMINO)PENTANEDIOIC ACID, 1-*tert*-BUTYL ESTER-5-METHYL ESTER: L-Glutamic acid, *N*-(diphenylmethylene)-, 1-(1,1-dimethylethyl)-5-methyl ester; (212121-62-5), **80**, 38
Benzil: Ethanedione, diphenyl-; (134-81-6) **82**, 10
rac-Benzoin: Ethanone, 2-hydroxy-1,2-diphenyl-; (19-53-9) **82**, 10
Benzonitrile; (100-47-0), **81**, 123
1,4-Benzoquinone: 2,5-Cyclohexadiene-1,4-dione; (106-51-4), **80**, 233; **82**, 1
Benzyl alcohol: Benzenemethanol; (100-51-6) **82**, 120
Benzylamine: Benzenemethanamine; (100-46-9), **81**, 262; **82**, 170
(*2S,6S*)-4-Benzyl-1,7-bis(trifluoromethylsulfonyl)-2,6-diisopropyl-1,4,7-triazaheptane: *N,N'*-[[(Phenylmethyl)imino]bis[(1*S*)-1-(1-methylethyl)-2,1-ethanediyl]]bis[1,1,1-trifluoro]-methanesulfonamide; (200351-80-0) **82**, 170
N-Benzylidene-*p*-anisidine: Benzenamine, 4-methoxy-*N*-(phenylmethylene)-; (783-08-4), **80**, 160
N-Benzylidene-benzylamine *N*-oxide; Benzenemethanamine, *N*-(phenylmethylene)-, *N*-oxide; (3376-26-9), **81**, 204
trans-2-BENZYL-1-METHYLCYCLOPROPAN-1-OL; 1-Cyclopropanol, 1-methyl-2-phenylmethyl-; **80**, 111
N-BENZYL-4-PHENYLBUTYRAMIDE: Benzenebutanamide, *N*-(phenylmethyl)-; (179923-27-4), **81**, 262
BICYCLO[2.2.1]HEPT-5-ENE-2-ENDO-CARBOXYLIC ACID, 3-ENDO-BENZYLOXYCARBONYL, (2*R*,3*S*)-: Bicyclo[2.2.1]hept-5-ene-2,3-dicarboxylic acid, mono(phenylmethyl) ester, (1*R*,2*S*,3*R*,4*S*)-; (581100-26-7) **82**, 120
endo-Bicyclo[2.2.1]hept-5-ene-2,3-dicarboxylic anhydride: 4,7-Methanoisobenzofuran-1,3-dione, 3a,4,7,7a-tetrahydro-, (3a*R*,4*S*,7*R*,7a*S*)-rel-; (129-64-6) **82**, 120
BICYCLO[3.1.0]HEXAN-1-OL; (7422-09-5), **80**, 111
(*R*)-BINAP: Phosphine, (1*R*)-[1,1: binaphthalene]-2,2'-diayhdris (diphenyl)-; (76189-55-4) **81**, 178
Bis(2-bromoethyl) ether: Ethane, 1,1'-oxybis[2-bromo-; (5414-19-7) **82**, 87
2,2-Bis(chloromethyl)-5,5-dimethyl-1,3-dioxane: 1,3-Dioxane, 2,2-bis(chloromethyl)-5,5-dimethyl-; (133961-12-3), **80**, 144
(*S,S*)-1,2-BIS-(*tert*-BUTYLMETHYLPHOSPHINO)ETHANE ((*S,S*)-*t*-Bu-BISP*)**82**, 22
Bis (tri-*tert*-butylphosphine)palladium: Palladium, bis[tris(1,1,-dimethylethyl)phosphine]- (9); (53199-31-8), **81**, 63
Bis(η^4-1,5-cyclooctadiene)-di-μ-methoxy-diiridium(I) ($[Ir(OMe)(COD)]_2$):

Carbonyldihydridotris(triphenylphosphine)ruthenium(II): Ruthenium, carbonyldihydridotris(triphenylphosphine); (25360-32-1), **80**, 104
Catalytic reduction **82**, 1
Cesium carbonate: Carbonic acid, dicesium salt; (534-17-8) **82**, 69
Cesium hydroxide: Cesium hydroxide (CsOH); (21351-79-1), **80**, 38
CHIRAL LEWIS ACID TRIDENTATE LIGAND, **80**, 46
Chloroacetic acid; (79-11-8) **82**, 157
Chloroacetyl chloride: Acetyl chloride, chloro-; (79-04-9), **80**, 200
Chlorobenzene: Benzene, chloro-(8,9); (108-90-7), **81**, 64
4-Chlorobenzoic acid methyl ester: Benzoic acid, 4-chloro-, methyl ester; (1126-46-1), **81**, 34
4-Chlorobenzonitrile: Benzonitrile, 4-chloro- (9); (623-03-0), **81**, 64
2-CHLORO-1,3-BIS(DIMETHYLAMINO)TRIMETHINIUM HEXAFLUOROPHOSPHATE: Methanaminium, *N*-[2-chloro-3-(dimethylamino)- 2-propenylidene]-N-methyl-, hexafluorophosphate(1-); (249561-98-6), **80**, 207
1-CHLORO-4-(2,2-DICHLORO-1-METHYLVINYL)BENZENE: Benzene, 1-chloro-4-(2,2-dichloro-1-methylethenyl)-; (73644-88-9) **82**, 93
1-Chloro-3-iodobenzene; (625-99-0) **82**, 126
1-CHLORO-3-IODO-5-(4,4,5,5-TETRAMETHYL-1,3,2-DIOXABOROLAN-2-YL)BENZENE: (2-(3-Chloro-5-iodophenyl)-4,4,5,5-tetramethyl-1,3,2-dioxaborolane; (479411-94-4) **82**, 126
m-Chloroperbenzoic acid; Benzenecarboperoxoic acid, 3-chloro-; (937-14-4), **80**, 207
9-Chloromethylanthracene: Anthracene, 9-(chloromethyl)-; (24463-19-2), **80**, 38
1-(4-Chlorophenyl)ethanone; (99-91-2) **82**, 93
1-(4-Chlorophenyl)ethanone hydrazone; (40137-41-5) **82**, 93
N-Chlorosuccinimide: 2,5-Pyrrolidinedione, 1-chloro-; (128-09-6), **80**, 133
Chlorotitanium triisopropoxide: Titanium, chlorotris(2-propanolato)-, (T-4)-; (20717-86-6), **80**, 111
Chlorotrimethylsilane: Silane, chlorotrimethyl- (9); (75-77-4) **80**, 172; **81**, 26, 216
Chromium(III) Cl Complex: Chromium, chloro[(1*R*,2*S*)-2,3-dihydro-1-[[[2-(hydroxy-κO)-5-methyl-3-tricyclo[3.3.1.13,7dec-1-ylphenyl]methyl-ene]amino-κN]-1H-indene-2-olato-(2-)-κO],(SP-4-4); (231963-76-1) **82**, 34
Cinchonidine: Cinchonan-9-ol, (8α, 9*R*)-; (485-71-2), **80**, 38
R-(+)-Citronellal: 6-Octenal, 3,7-dimethyl-, (3*R*)-; (2385-77-5), **80**, 195
Claisen condensation, **81**, 178
(*S*)-COP-Cl: Cobalt, bis[1,1',1",1'"-(η4-1,3-cyclobutadiene-1,2,3,4-tetrayl)tetrakis[benzene]](di-μ-chlorodipalladium)bis[μ-[(1-η:1,2,3,4,5-η)-2-[(4*S*)-4,5-dihydro-4-(1-methylethyl)-2-oxazolyl-kN3]-2,4-cyclopentadien-1-yl]]di-; (581093-92-7) **82**, 134
Copper(II) acetate monohydrate: Acetic acid, copper(2+) salt, monohydrate; (6046-93-1) **82**, 108
Copper (I) chloride: Copper chloride: (7758-89-6) **82**, 69, 93
Copper (II) chloride: Copper chloride; (7447-39-4) **82**, 22
Copper(I) Iodide: Copper iodide (CuI); (7681-65-4), **80**, 129
Copper-mediated phenol coupling **82**, 69
Cross coupling, **81**, 33

Furfural: 2-Furancarboxaldehyde; (98-01-1), **80**, 66

Geranial: 2,6-Octadienal, 3,7-dimethyl-, (2*E*)-; (141-27-5) **82**, 80
Geraniol: 2,6-Octadien-1-ol, 3,7-dimethyl-, (2*E*)-; (106-24-1) **82**, 80
(*R*)-(–)-Glycidyl butyrate: Butanoic acid, (2*R*)-oxiranylmethyl ester; (60456-26-0), **81**, 112
Glycosylation, **81**, 225
Grignard reagents, **81**, 33
Grubbs catalyst, **80**, 85; **81**, 4

Heck reaction, **81**, 63
HELICENEBISQUINONES, **80**, 233
3-Heptanol; (589-82-2), **81**, 244
(*Z*)- 1-HEPTENYLDIMETHYLSILANOL: Silanol, (1*Z*)-1-heptenyldimethyl-; (261717-40-2), **81**, 42
(Z)-1-HEPTENYL-4-METHOXYBENZENE: BENZENE, 1-(1*Z*)-1-HEPTENYL-4-METHOXY-; (80638-85-3), **81**, 42, 54
3-Heptyl benzenesulfenate: Benzenesulfenic acid, 1-ethylpentyl ester; (198778-69-7), **81**, 244
1-Heptyne: 1-Heptyne; (628-71-7), **81**, 42, 54
Hetero-Diels Alder **82**, 34
Hexabutylditin: Distannane, hexabutyl; (813-19-4), **81**, 245
Hexacarbonyl[μ[(3,4-η:3,4-η)-2-methyl-3-butyn-2-ol]]dicobalt: Cobalt, hexacarbonyl[μ-[(3,4-η:3,4-η)-2-methyl-3-butyn-2-ol]]di-, (Co-Co); (40754-33-4), **80**, 93
Hexafluorophosphoric acid: Phosphate (1-), hexafluoro-, hydrogen; (16940-81-1), **80**, 200
HEXAKIS(4-BROMOPHENYL)BENZENE (HBB) **82**, 30
hexakis(4-Bromophenyl)benzene: 1,1':2',1''-Terphenyl, 4,4''-dibromo-3',4',5',6'-tetrakis(4-bromophenyl)-; (19057-50-2) **82**, 30
Hexamethylcyclotrisiloxane: Cyclotrisiloxane, hexamethyl-; (54-05-9), **81**, 44
1,1,1,3,3,3-Hexamethyldisilazane: Silanamine,1,1,1-trimethyl-N-(trimethylsilyl)-; (999-97-3), **80**, 160; **82**, 140, 147
Hexaphenylbenzene: 1,1':2',1''-Terphenyl, 3',4',5',6'-tetraphenyl-; (992-04-1) **82**, 30
1-Hexene, 6-iodo-; (18922-04-8), **81**, 121
trans-2-Hexen-1-ol: 2-Hexen-1-ol, (2*E*)-; (928-95-0) **82**, 134
5-Hexen-1-ol; (821-41-0). **81**, 121
5-Hexen-1-ol, methanesulfonate; (64818-36-6), **81**, 121
Hydantoin synthesis, **81**, 213
Hydrazine; (10217-52-4), **81**, 254
Hydrazine hydrate: Hydrazine, monohydrate; (7803-57-8) **82**, 93
Hydriodic acid; (10034-85-2), **80**, 129
(*R*,*R*)-Hydrobenzoin: 1,2-Ethanediol, 1,2-diphenyl-, (1*R*,2*R*)-; (52340-78-0) **82**, 10
Hydrocinnamaldehyde: Benzenepropanal; (104-53-0) **82**, 170
Hydrogen peroxide (H_2O_2); (7722-84-1), **80**, 9, 184; **82**, 80, 157

N-Phenylcarbamic acid methyl ester: Carbamic acid, phenyl-methyl ester; (2603-10-3), **81**, 112
1-Phenylcyclohexene: Benzene, 1-cyclohexen-1-yl-; (771-98-2), **80**, 9
(*R,R*)-1-PHENYLCYCLOHEXENE OXIDE: 7-Oxabicyclo[4.1.0]heptane, 1-phenyl-, (1*R*,6*R*)-; (17540-04-4), **80**, 9
S-Phenyl diazothioacetate: Ethanethioic acid, diazo-, *S*-phenyl ester; (72228-26-3), **80**, 160
Phenyldimethylsilane: Silane, dimethylphenyl-; (766-77-8) **82**, 188
(*E*)-4-(2-PHENYLETHENYL)BENZONITRILE: BENZONITRILE, 4-[(1*E*)-2-PHENYLETHENYL]- (9); (13041-79-7), **81**, 63
(*S*)-1-Phenylethylamine: Benzenemethanamine, α-methyl-, (α*S*)-; (2627-86-3), **80**, 207
(*S*)-[(1-Phenylethyl)amino]acetonitrile; Acetonitrile, [[(1*S*)-1-phenylethyl]amino]-; (35341-76-5), **80**, 207
[(1*S*)-1-Phenylethyl]imino]acetonitrile *N*-oxide; Acetonitrile, [oxido[(1*S*)-1-phenylethyl]imino]-; (300843-73-6), **80**, 207
(4*S*)-4-(2-PHENYLETHYL)-2-OXETANONE: (4*S*)-4-(2-phenylethyl)-2-oxetanone; (214853-90-4) **82**, 170
(*R*)-(−)-2-Phenylglycinol: Benzeneethanol, β-amino-, (β*R*)-; (56613-80-0), **80**, 46
N-PHENYL-5R-HYDROXYMETHYL-2-OXAZOLIDINONE: 2-Oxazolidinone, 5-(hydroxymethyl)-3-phenyl-, (5*R*); (875080-42-7), **81**, 112
Phenylmagnesium bromide: Magnesium, bromophenyl-; (100-58-3), **80**, 57
(*R*)-4-Phenyl-2-oxazolidinone: 2-Oxazolidinone, 4-phenyl-, (4*R*)-; (90319-52-1), **81**, 147
5-PHENYLPENT-2-YN-1-OL: 2-Pentyn-1-ol, 5-phenyl-; (16900-77-9), **81**, 2
R-4-PHENYL-3-(1,2-PROPADIENY)-2-OXAZOLIDINONE: 2-OXAZOLIDINONE, 4-PHENYL-3-(1,2-PROPADIENYL)-, (4*R*)-; (256382-50-0), **81**, 147
3-Phenylpropionaldehyde: Benzenepropanal; (104-53-0), **81**, 2
3-Phenyl-2-propylthio-2-propen-1-ol, **80**, 190
3-Phenyl-2-propyn-1-ol: 2-Propyn-1-ol, 3-phenyl-; (1504-58-1), **80**, 190
R-4-Phenyl-3-(2-propynyl)-2-oxazolidinone: 2-Oxazolidinone, 4-phenyl-3-(2-propynyl); (4*R*)-; (256382-74-8), **81**, 147
S-Phenyl thioacetate: Ethanethioic acid, *S*-phenyl ester; (934-87-2), **80**, 160
2-PHENYLTHIO-5-HEPTANOL: 3-Heptanol, 6-(phenylthio); (198778-75-5), **81**, 244
Phosphorus oxychloride: Phophoric trichloride; (10025-87-3), **80**, 200
Photolysis, **81**, 245
Pinacol: 2,3-Butanediol, 2,3-dimethyl-; (76-09-5), **81**, 89
Pinacolborane: 4,4,5,5-Tetramethyl-1,3,2-dioxaborolane; (25015-63-8) **82**, 126
Pinacolic coupling, **81**, 26
Piperidine-4-spiro-5'-hydantoin: 1,3,8-Triazaspiro [4.5]decane-2,4-dione; (13625-39-3), **81**, 214
4-Piperidone monohydrate hydrochloride: 4-Piperidinone, hydrochloride; (41979-39-9), **81**, 214
Pivaloyl chloride: Propanoyl chloride, 2,2-dimethyl-; (3282-30-2), **81**, 226, 254
Platinum(0)-1,3-divinyl-1,1,3,3-tetramethyldisiloxane complex: Platinum, 1,3-diethenyl-1,1,3,3-tetramethyldisiloxane complex; (68478-92-2), **81**, 55

Rose Bengal; (11121-48-5), **80**, 66
Ruthenium, dodecacarbonyltri-, triangulo; (15243-33-1) **82**, 188
Ruthenium-mediated reduction **82**, 10
RuCl[(1*S*,2*S*)-p-TsNCH(C_6H_5)CH(C_6H_5)NH_2](η^6-p-cymene): Ruthenium, [*N*-[(*1S,2S*)-2-(amino-κN)-1,2-diphenylethyl]-4-methyl-benzenesulfonamidato-κN]chloro-[(1,2,3,4,5,6-η)-1-methyl-4-(1-methylethyl)benzene]-; (192139-90-5) **82**, 10

Salicyl alcohol: Benzenemethanol, 2-hydroxy-; (90-01-7), **81**, 171
Silver carbonate: Carbonic acid, disilver(1+) salt; (534-16-7) **82**, 75
Silylation of alcohols, **81**, 157
Sodium; (7440-23-5), **80**, 144
Sodium amide: Sodium amide ($NaNH_2$); (7782-92-5), **80**, 144; **82**, 140
Sodium bicarbonate: Carbonic acid monosodium salt; (144-55-8) **82**, 75
Sodium bisulfite: Sulfurous acid, monosodium salt; (7631-90-5) **82**, 30, 75
Sodium carbonate: Carbonic acid disodium salt; (497-19-8) **82**, 55
Sodium chlorite: Chlorous acid, sodium salt; (7758-19-2), **81**, 195
Sodium dithionite: Dithionous acid, disodium salt; (7775-14-6), **80**, 227
Sodium fluoride: Sodium fluoride (NaF); (7681-49-4), **80**, 172
Sodium hydride: Sodium hydride (NaH); (7646-69-7), **81**, 147
Sodium hydrogen sulfite: Sulfurous acid, monosodium salt: (7631-90-5) **82**, 1
Sodium hypochlorite: Hypochlorous acid, sodium salt (8,9); (7681-52-9), **81**, 195; **82**, 80
Sodium iodide: Sodium iodide (NaI); 7681-82-5, **81**, 77, 121
Sodium metaperiodate: Periodic acid (HIO_4), sodium salt; (7790-28-5), **81**, 171
Sodium methoxide: Methanol, sodium salt; (124-41-4), **80**, 133
Sodium nitrite: Nitrous acid, sodium salt; (7632-00-0), **80**, 133
Sodium tetrafluoroborate; (13755-29-8) **82**, 166
Sodium thiosulfate: Thiosulfuric acid ($H_2S_2O_3$), disodium salt; (7772-98-7), **81**, 78; **82**, 157
Solid supports, **81**, 235
(-)-Sparteine: 7,14-Methano-2H,6H-dipyrido[1,2-a:1',2'-e][1,5]diazocine, dodecahydro-, (7*S*,7a*R*,14*S*,14a*S*)-; (90-39-1) **82**, 22
9-SPIROEPOXY-endo-TRICYCLO[5.2.2.$0^{2,6}$]UNDECA-4,10-DIEN-8-ONE: Spiro[4,7-ethano-1H-indene-8,2'-oxiran]-9-one, 3a,4,7,7a-tetrahydro-; (146924-02-9), **81**, 171
Styrene: Benzene, ethenyl-; (100-42-5), **81**, 65
Sulfenate synthesis, **81**, 244
Sulfide synthesis, **81**, 244
Sulfuryl chloride; (7791-25-5), **81**, 247
Suzuki synthesis, **81**, 89

TETRABENZYL PYROPHOSPHATE: Diphosphoric acid, tetrakis(phenylmethyl) ester; (990-91-0), **80**, 219
Tetrabutylammonium bromide: 1-Butanaminium, *N,N,N*-tributyl-, bromide; (1643-19-2), **80**, 227

Tetrabutylammonium fluoride trihydrate: 1-Butanaminium, *N, N, N*-tributyl, fluoride, trihydrate; (87749-50-6), **81**, 45, 55
Tetrabutylammonium hydrogen sulfate: 1-Butanaminium, *N,N,N*-tributyl-, sulfate (1:1); (32503-27-8), **80**, 9
Tetrachloromethane: Methane, tetrachloro-; (56-23-5) **82**, 18
Tetrafluoroboric acid: Borate(1-), tetrafluoro-, hydrogen (8,9); (16872-11-0) **82**, 80
2,3,3α,4-TETRAHYDRO-2-[(4-METHYLBENZENE)SULFONYL]CYCLOPENTA-[C]PYRROL-5(1H)-ONE: Cyclopenta[b]pyrrol-5(1h)-one, 2,3,3a,4-tetrahydro-1-[(4-methylphenyl)sulfonyl]-; (205885-50-3), **80**, 93
Tetrakis(hydroxymethyl)phosphonium sulfate ("Pyroset-TKOW"): Phosphonium, tetrakis(hydroxymethyl)-, sulfate (2:1); (55566-30-8), **80**, 85
1-Tetralone: 1(2H)-Naphthalenone, 3,4-dihydro-; (529-34-0), **80**, 104
1,1,4,4-Tetramethoxybut-2-yne: 2-Butyne, 1,1,4,4-tetramethoxy-; (53281-53-1) **82**, 179
2-(4,4,5,5,-TETRAMETHYL-1,3,2-DIOXABOROLAN-2-YL)INDOLE: 2-(4,4,5,5-Tetramethyl-1,3,2-dioxaborolan-2-yl)-1H-indole; (476004-81-6) **82**, 126
3-(4,4,5,5-Tetramethyl-[1,3,2]dioxaborolan-2-yl)-pyridine: Pyridine, 3-(4,4,5,5-tetramethyl-1,3,2-dioxaborolan-2-yl)-; (329214-79-1), **81**, 89
1,1,3,3,-Tetramethyldisiloxane: Disiloxane, 1,1,3,3,-tetramethyl-; (3277-26-7), **81**, 54
N,N,N',N'-Tetramethylethylenediamine: 1,2-Ethanediamine, *N,N,N',N'*-tetramethyl-; (110-18-9), **80**, 46
2,2,6,6-Tetramethylheptane-3,5-dione: 3,5-Heptanedione, 2,2,6,6-tetramethyl-: (1118-71-4) **82**, 69
2,2,6,6-Tetramethylpiperidine; Piperidine, 2,2,6,6-tetramethyl-; (768-66-1), **81**, 134
2,2,6,6-Tetramethyl-1-piperidinyloxy (TEMPO): 1-Piperidinyoxy, 2,2,6,6-tetramethyl-; (2564-83-2), **81**, 195
"Thia-Wolff" rearrangement, **80**, 166
Thioanisole: Benzene, (methylthio)-; (100-68-5), **80**, 184
Thiophenol: Benzenethiol; (108-98-5), **81**, 246
Thionyl chloride; (7719-09-7), **80**, 46
Titanium tetrachloride: Titanium chloride ($TiCl_4$)(T-4); (7550-45-0), **81**, 16
Titanium tetraisoproproxide: 2-Propanol, titanium (4+)salt; (546-68-9) **80**, 120; **81**, 16
Toluene: Benzene, methyl-; (108-88-3) **82**, 55
p-Toluenesulfonyl chloride: Benzenesulfonyl chloride, 4-methyl-; (98-59-9) **80**, 93, 133; **81**, 77, 159
5-(Tosyloxyimino)-2,2-dimethyl-1,3-dioxane-4,6-dione: 1,3-Dioxane-4,5,6-trione, 2,2-dimethyl-, 5-O-[(4-methylphenyl)sulfonyl]oxime; (215436-24-1), **80**, 133
3,4,6-Tri-*O*-benzyl-D-glucal: D-*arabino*-Hex-1-enitol, 1,5-anhydro-2-deoxy-3,4,6-tris-*O*-(phenylmethyl)-; (55628-54-1), **81**, 225
3,4,6-TRI-*O*-BENZYL-2-*O*-PIVALOYL-β-D-GLUCOPYRANOSYL-(1→6)-1,2:3,4-DI-*O*-ISOPROPYLIDENE- α-D-GALACTOPYRANOSIDE: α-D-Galactopyranose, 6-O-[2-O-(2,2-dimethyl-1-oxopropyl)-3,4,6-tris-*O*-(phenylmethyl)-β-D-glucopyranosyl}-1,2:3,4-bis-*O*-(1-methylethylidene)-; [219122-26-6], **81**, 226